THE MUHU PROCEEDINGS 2

ISOLATED ISLANDS IN MEDIEVAL NATURE, CULTURE AND MIND

CEU MEDIEVALIA 14

Series Editor: József Laszlovszky
Series Technical Editor: Annabella Pál

Isolated Islands in Medieval Nature, Culture and Mind

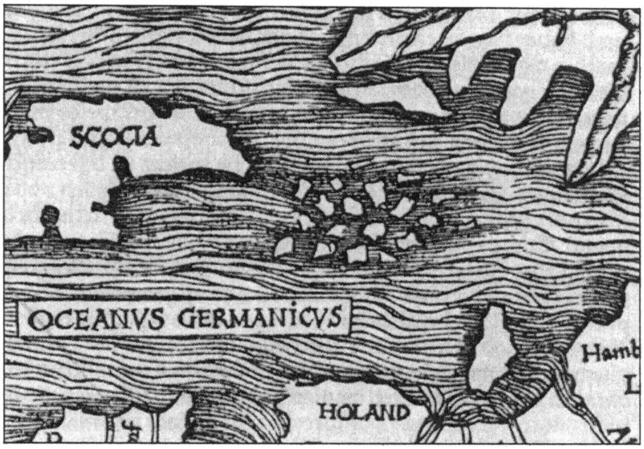

Edited by
Torstein Jørgensen
and
Gerhard Jaritz

Centre for Medieval Studies
University of Bergen
&
Central European University
Department of Medieval Studies
&
Central European University Press
Budapest · New York

Budapest, 2011

1st edition

Technical Editor: Gerhard Jaritz
Copy Editor: Judith Rasson
Cover design for the series by Péter Tóth

Cover Illustration:
Islands in the North of Europe
Hartmann Schedel, World Chronicle (Nuremberg, 1493), fol. 286v (detail)

Joint publication by:

Centre for Medieval Studies, University of Bergen
P.O.Box 7805, 5020 Bergen, Norway
Telephone: +47 55 58 80 85, Fax: +47 55 58 80 90
E-mail: post@cms.uib.no, Website: http://www.uib.no/cms/en

Central European University
Department of Medieval Studies
Nádor u. 9, H-1051 Budapest, Hungary
Telephone: (+36-1) 327-3051, Fax: (+36-1) 327-3055
E-mail: medstud@ceu.hu, Website: http://medstud.ceu.hu

Central European University Press
An imprint of the Central European University Share Company
Nádor u. 11, H-1051 Budapest, Hungary
Telephone: (+36-1) 327-3138, Fax: (+36-1) 327-3183
E-mail: ceupress@ceu.hu, Website: http://www.ceupress.com

400 West 59th Street, New York NY 10019, USA
Telephone (+1-212) 547-6932, Fax: (+1-646) 557-2416
E-mail: mgreenwald@sorosny.org

Published with the support of NCMS

ISSN 1587-6470 CEU MEDIEVALIA
ISBN 978-615-5053-24-5

Library of Congress Cataloging-in-Publication Data

Isolated Islands in Medieval Nature, Culture and Mind / edited by Torstein Jørgensen, Gerhard Jaritz – 1st ed.
p. cm. -- (CEU medievalia ; 14)
Includes bibliographical references and index.
ISBN 978-6155053245 (pbk.)
1. Islands--History. 2. Islands--Social aspects. 3. Civilization, Medieval.
I. Jørgensen, Torstein, 1951- II. Jaritz, Gerhard, 1949-
GB471.I88 2011
304.2'3--dc22

2010052374

Printed in Hungary by Akaprint Kft., Budapest

TABLE OF CONTENTS

List of Illustrations vi

Preface 1

Felicitas Schmieder, Paradise Islands in the East and West – Tradition and Meaning in Some Cartographical Places on the Medieval Rim of the World 3

Kristel Zilmer, The Powers and Purposes of an Insular Setting – on Some Motifs in Old-Norse Literature 23

Johnny Grandjean Gøgsig Jacobsen, Monastic "Islands" in Medieval Denmark:: Insular Isolation in Ideal and Practice 36

Torben Kjersgaard Nielsen, Religious Athletes – on the Perception of the Body in Medieval Asceticism 45

Eldar Heide, Holy Islands and the Otherworld: Places beyond Water 57

Else Mundal, Hvítramannaland and Other Fictional Islands in the Sea 81

Margaret Elphinstone, The Unknown Island 88

Bernt Øyvind Thorvaldsen, Will the Son of Nine Sisters Rule the Sea-Kidney? 97

Gerhard Jaritz, The Quality of Islands in Middle High German Literature 111

Juhan Kreem, Seasonal Isolation in the Communication in Livonia 120

Torstein Jørgensen, Utstein Monastery: An Island on an Island – or not? 128

List of Contributors 139

Index 140

LIST OF ILLUSTRATIONS

Felicitas Schmieder, "Paradise Islands in the East and West"
p. 4, fig. 1: Paradise on an island at the Eastern end of the world, late 12th century
p. 5, fig. 2: Paradise in *Rudimentum noviciorum epitome*, 1475
p. 5, fig. 3: Paradise in the East on an island/peninsula, giving birth to the Four Rivers, 12th century
p. 6, fig. 4: Paradise on an island in the East, 13th century
p. 8, fig. 5: Paradise in East Africa close to Prester John, c. 1450
p. 10, fig. 6: Paradise as *locus deliciarum* on the Velletri/Borgia map, 15th century
p. 13, fig. 7: Atlas Catalan, 1375
p. 14, fig. 8: Canary islands (*fortunatum insule*) and *illa de brezill*, c. 1450
p. 15, fig. 9: Andrea Bianco, *Atlante nautico*, 1436
p. 18, fig. 10: Island of *brezill* on the Catalan Estense world map, c. 1450
p. 19, fig. 11: Andrea Bianco, *Atlante nautico*, 1436
p. 21, fig. 12: Andreas Walsperger, World map, 1448
p. 22, fig. 13: Detail from the dedication page of Jean Germain, *Mappemonde spirituelle*, 1449

Johnny Grandjean Gøgsig Jacobsen, "Monastic "Islands" in Medieval Denmark"
p. 39, fig. 1: Distribution of monasteries in medieval Denmark
p. 43, fig. 2: Reconstructed city map of Århus showing the location of the Dominican priory

Juhan Kreem, "Seasonal Isolation in the Communication in Livonia"
p. 119, fig. 1: Medieval Livonia and its neighbours

PREFACE

An international network of scholars has over the last years been focusing on matters related to the peripheral areas and outer borders of the medieval world. The overarching perspective has been comprehensive in the sense that the idea of periphery, liminality, borders and outskirts has been understood and treated in a wide sense. Thus, the network has included scholars and contributions dealing with these aspects under a wide range of angles – such as geography and mapping, fantasy and fiction, travelling, social life of somehow marginalised people, religion, time, and mind and body.

The network is also inclusive as far as membership is concerned, and participation changes depending on what main aspect is in focus at different times. So far the network has been administered by Gerhard Jaritz (Budapest and Krems), Juhan Kreem (Tallinn), and Torstein Jørgensen (Bergen and Stavanger), and has started a tradition of convening for workshops and exchange of viewpoints and findings every second year.

The network is publishing its results in the series of The Muhu Proceedings, named after the first venue of the group on the Estonian island of Muhu. The first volume was published in 2009. It concentrated on the medieval discourses about the edges of the world.[1]

In this second volume we have narrowed the scope to focus more specifically on the concept and role of islands in the medieval world. The main characteristic of an island is, of course, that of being isolated from the rest of the world; in geography by waters, in more abstract and symbolic meanings by other kinds of separating borders. Islands were the place 'on the other side', of difference, otherness and remoteness. As one of the articles in this volume puts it, islands are often depicted "as sites for extraordinary events and happenings".

The eleven selected contributions to this book cluster around the concept of islands under the following six perspectives:

- the search for undiscovered and paradise islands in literature and cartography (M. Elphinstone, F. Schmieder);
- the idea of fictional islands in Old Norse texts (E. Heide, E. Mundal, B. Ø. Thorvaldsen, K. Zilmer);

---

1 Gerhard Jaritz and Juhan Kreem (ed.), *The Edges of the Medieval World*, CEU Medievalia 11 (Budapest and New York:: Central European University Press, 2009).

- the quality of islands in literary evidence (G. Jaritz);
- monasteries as islands and monasteries on islands (J. G. G. Jakobsen, T. Jørgensen);
- the human body as an island of religious achievements (T. K. Nielsen);
- seasonal islandisation of certain lands (J. Kreem).

November, 2010

Torstein Jørgensen (Bergen and Stavanger)
Gerhard Jaritz (Budapest and Krems)

PARADISE ISLANDS IN THE EAST AND WEST – TRADITION AND MEANING IN SOME CARTOGRAPHICAL PLACES ON THE MEDIEVAL RIM OF THE WORLD

Felicitas Schmieder

Paradise was, in Christian medieval thinking, a special and meaningful place, both in time and in space: It had been there since the very beginning of history when God created it as a home for the first human beings – and would still be there at history's very end. Revelations 2:7 alludes to its role at the end of all times, promising

> He that hath an ear, let him hear what the Spirit saith unto the churches; To him that overcometh will I give to eat of the tree of life, which is in the midst of the paradise of God.

Considering this, paradise had to exist through all time, thus also in present – somewhere on earth, but unreachably. The Franciscan, John of Marignola, who travelled to East Asia and also to Indian islands (present-day Indonesia) in the mid-fourteenth century, tells that he could only hear the waters of paradise island rush in the distance. It was not a wide step to imagine a place secluded from the rest of the world not on the margins of the inhabited earth, surrounded by high mountains, but as an island, or, as John of Marignola puts it: *Est autem paradisus locus in terra circumvallatus mari oceano in parte orientali* (Paradise is a place on earth, surrounded by the ocean in the East).[1]

Consequently, this imagining happened frequently. We can trace it on late medieval world maps.[2] These *mappae mundi* picture paradise quite regularly, named or at least easily recognizable, with clear features such as the four rivers of paradise (which John of Marignola heard), the tree (sometimes with the snake), and Adam and Eve, not always but often on an island or at least a peninsula at the very rim of the earth. On the more historically oriented maps (those dealing with human actions) one can find not only

---

1 Johannes de Marignola, *Relatio* (= excerpts concerning his voyage to Asia from his *Cronicon Boemorum*), in *Sinica Franciscana I: Itinera et Relationes Fratrum Minorum saeculi XIII et XIV*, ed. Anastasius van den Wyngaert (Quaracchi: Collegium S. Bonaventurae, 1929), 531.

2 For the representation of paradise on medieval maps in a broader context cf. Alessandro Scafi, *Mapping Paradise: A History of Heaven on Earth* (London: British Library, 2006).

Adam and Eve as signatures for paradise, but a depiction of them being driven out of paradise by the Archangel Michael (fig. 1-4).[3]

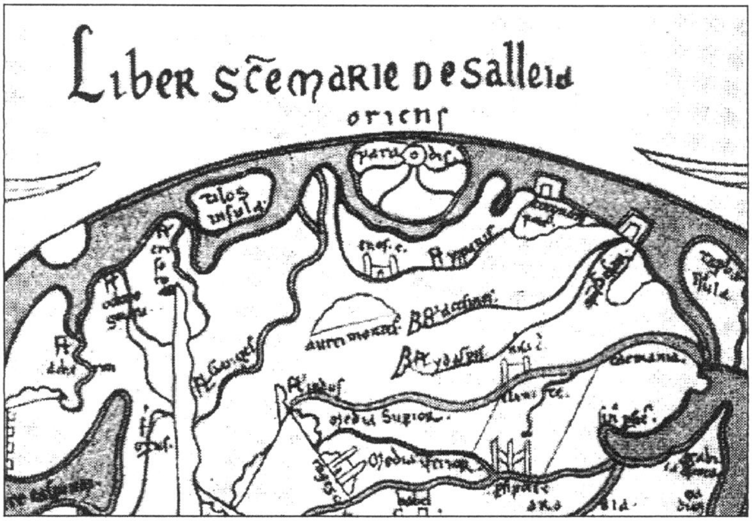

Fig. 1: Paradise on an island at the Eastern end of the world with the Four Rivers on an oval world map (29,5 x 20,5 cm), late 12th c., from Ms. Cambridge Corpus Christi Cod. LXVI, p.2 (known as the map by Henry of Mainz; nowadays frequently called Sawley world map (from Hahn-Woernle, *Ebstorfer Weltkarte*, cf. n. 3, p. 31)

---

3 The examples in this article are taken from several *mappae mundi* created between 1200 and 1500: world map in Lambert of Saint-Omer, *Liber Floridus*, 12th c., http://diglib.hab.de/wdb.php?dir=mss/1-gud-lat; – world map from Ms. Cambridge Corpus Christi Cod. LXVI, p.2; – the Hereford Map, 13th c.: P. D. A. Harvey (ed.), *Mappa Mundi. The Hereford Map* (Toronto: University of Toronto Press, 1996); cf. Scott D. Westrem (ed.), *The Hereford Map* (Turnhout: Brepols, 2001); – Atlas Catalan, 1375: *Atlas Catalan von 1375*, ed. and trans. Hans-Christian Freiesleben (Stuttgart: Brockhaus, 1977); – Velletri or Borgia World Map, 15th c.: Nils Adolf Erik Nordenskiöld (ed.), Periplus. *An Essay on the Early History of Charts and Sailing Directions* (Stockholm: Norstedt, 1897), plate XXXIX; – Andrea Bianco, *Atlante nautico* (including his World Map), 1436: ed.Piero Falchetta (Venice: Arsenale Ed., 1993); – *Mappa mundi* by Andreas Walsperger, 1448, see *Biblioteca Palatina. Katalog zur Ausstellung vom 8. Juli - 2. November 1986*, ed. Elmar Mittler et al. (Heidelberg: Edition Braus, 1986), 236; also in Guglielmo Cavallo (ed.), *Cristoforo Colombo e l'apertura degli spazi. Mostra storico-cartografica,* 2 vols. (Rome: Libreria dello stato, 1992), I, 75 – world map from the Biblioteca Estense, Modena, c. 1450: *Il Mappamondo Catalano Estense (Die Katalanische Estense Weltkarte)*, ed. E. Milano and A. Battini (Zurich: Urs-Graf-Verlag, 1995); – *Rudimentum noviciorum epitome sive systema historiae universalis auctore incerto confecta et in sex aetates divisa ab OC–1473*, 2 vols. in quart, Lübeck, 1475, fol.74v/75r. In recent years, several general works on old maps and cartography have started to gradually replace the classic work by Konrad Miller [*Mappae Mundi: die ältesten Weltkarten*, 6 vols. (Stuttgart: Roth, 1895-98)]: John B. Harley and David Woodward (ed.), *The History of Cartography*, vol 1: *Cartography in Prehistoric, Ancient, and Medieval Europe and the Mediterranean* (Chicago and London: University of Chicago Press, 1987); Evelyn Edson, *The World Map, 1300-1492: The Persistence of Tradition and Transformation* (Baltimore: Johns Hopkins University Press, 2007).

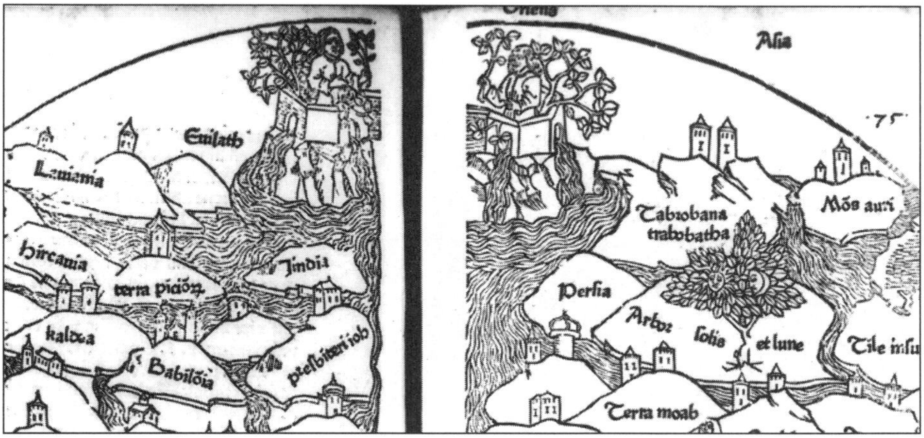

Fig. 2: *Rudimentum noviciorum epitome sive systema historiae universalis auctore incerto confecta et in sex aetates divisa ab OC–1473*, 2 vols. in quart, Lübeck 1475, fol.74v/75r (from Universitätsbibliothek Frankfurt am Main Inc.fol.118).

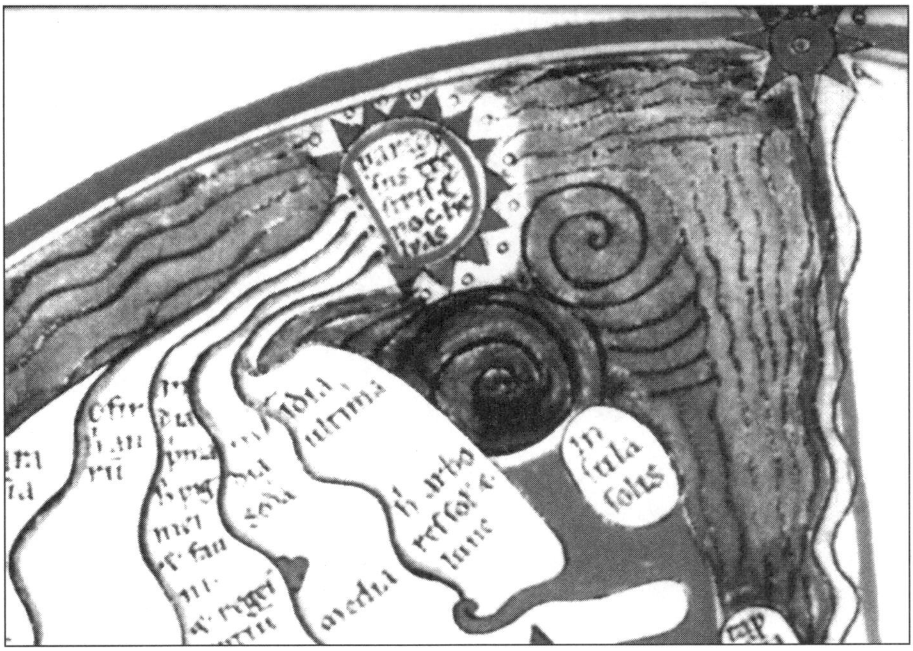

Fig. 3: Paradise in the East on an island/peninsula, giving birth to the Four Rivers, in Lambert of Saint-Omer, Liber Floridus (Ø c. 15 cm), 12th c., Ms. Herzog August-Bibliothek Wolfenbüttel Cod. Guelf. 1 Gud. lat. fol. 69v (from http://diglib.hab.de/mss/1-gud-lat/start.htm?image=00144).

Mostly, paradise on earth was represented in the East of the world, as not only Marignola says, but as Isidor of Seville had already made clear: *Paradisus est locus in orientis partibus constitutus.*[4] But this could change in certain circumstances; during the later Middle Ages, the Latin Europeans got used to adapting their geographical views to new findings and insights, as well as needs, hopes, and expectations. One of the most obvious examples of this adaptation is the following.

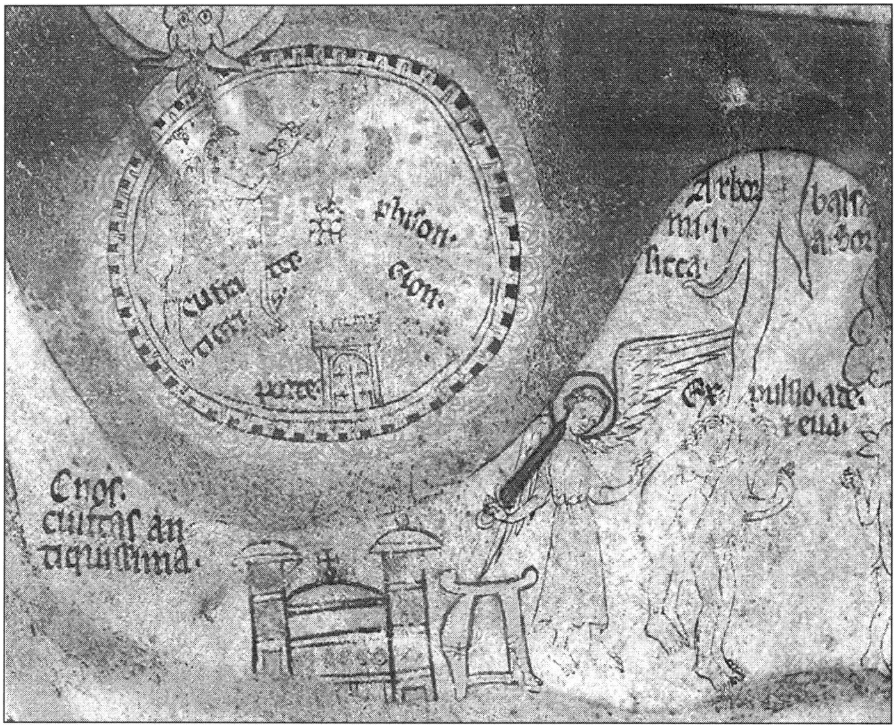

Fig. 4: Paradise on an island in the East, on Hereford world map (1,6 x 1,3 m, Ø 1,32 m), 13th c. (from Harvey, *Mappa Mundi*, cf. n. 3, p.40)

The four major rivers that were known as having their source in paradise and from there running through inhabited land were mostly imagined in Asia, namely, the Euphrates and Tigris, Phison (in the Later Middle Ages identified by some as the Ganges), and Gyon. The latter was normally identified as the Nile. Since the Nile River was considered one of the rivers that sprang from paradise and it seemed to come from

[4] Isidor Hispalensis Episcopi, *Etymologiarum sive origium libri* XX, ed. W. M. Lindsay (Oxford: Clarendon Press, 1911), XIV, 3, 2 = t. II, p.112: *Paradisus est locus in orientis partibus constitutus cuius vocabulum ex Graeco in Latinum vertitur hortus, porro Hebraice Eden dicitur quod in nostra lingua deliciae interpretatur.*

quite a different direction than the others, it was (as soon as Europeans reached the Middle Eastern regions as far as the Indian Ocean) a problem to explain it rising in the East and then running to where it was known to run.[5] When, during the fifteenth century, the idea of finding the source of the Nile and diverting it from Egypt became more and more interesting (in order to let the land of the Mamluks fall dry and thus weaken them as enemies in an upcoming crusade), the attempt shown on the world map from the Estense Library in Modena seems to have been a better localisation for the Catalan interests, especially at the time it was created (c. 1450;[6] fig. 5). It locates paradise in East Africa – and since eastern Africa was still considered part of India (known, like Paradise, as being far in the East of the world) anyway, the delocalisation was clear only on the maps, but not in principle.

While paradise in this case was not located on an island of useful adaptation, the island localisation could at the same time be strengthened by other traditions strongly present in the Middle Ages besides the biblical one. There had been ideas of a paradise and especially of paradisiacal islands from antiquity and early medieval legends of not-always-clear origin, but these places of paradise were not necessarily found in the East.

---

5 For the attraction of the idea of finding paradise at the sources of the Nile or in Ethiopia (wherever this country may have been thought to lie), see Henri Baudet, *Paradise on Earth. Some Thoughts on European Images of Non-European Man*, tr. E. Wentholt (New Haven and London: Yale University Press, 1965), esp. 15-22. Scafi, *Mapping* mentions at length the replacing of paradise on the Modena map and gives as a reason the growing importance of trade interests. I would like to stress here that it was not only trade, but politics based on trading interests as well as crusade hopes strengthened by eschatology.

6 The latest book re-evaluating the available evidence on the late medieval *mappae mundi* (Edson, *World Maps*, 197) notes 1446 as a clear *terminus post quem*. This puts it in the time of the Catalan-Aragonese King Alfons V, who was in close marital connection with the Portuguese king – which means that presumably new evidence from Portuguese discoveries on the west coast of Africa would have reached the mapmaker quite quickly. This validates the date of "c. 1450". A comparable "shifting" of paradise to a place more useful in present political aims and hopes was made by Francesco Roselli on his world map of 1507/08, where a paradisiacal place in the inland area of South America is not named as terrestrial paradise, but described or at least suggested as such by four rivers springing from it: cf. Francesca Fiorani, "The Places of Renaissance Mapping," in *Herrschaft verorten. Politische Kartographie des Mittelalters und der Frühen Neuzeit*, ed. Ingrid Baumgärtner and Martina Stercken (forthcoming). – There is clear evidence that the mapmakers knew about each others' products and sometimes they even "discussed" with each other. So, the entry in the "Genoese World Map" of 1457 – *In hac regione depinxerunt quidem paradisum deliciarum. Alii vero ultra Indias ad orientem eum esse dixerunt* [Edward L. Stevensom, ed., Genoese World Map 1457 (New York: Hispanic Society of America, 1912)] – may be a comment on the opinion expressed on the Modena Map.

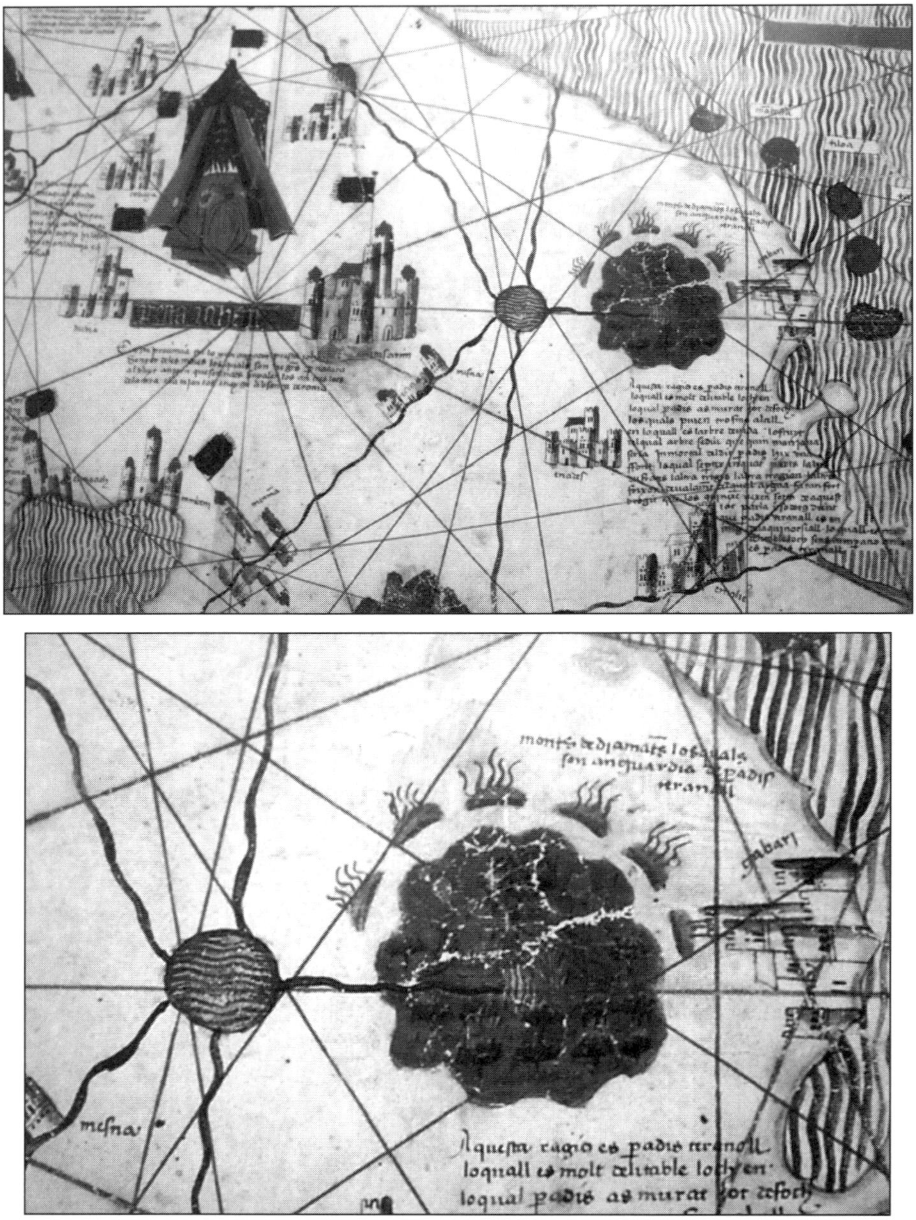

Fig. 5: Paradise in East Africa close to Prester John on the Catalan Estense world map at Modena (Ø 1,13 m) c. 1450 (from *Mappamondo Catalano Estense*, cf. n. 3).

At this point, I have to insert a remark on the terms and concepts I am dealing with. There are quite a few traditions of islands and other places easily described as paradisiacal which are not necessarily called that way in the sources, but which can be compared to or identified with paradise by certain features such as copious abundance, fecundity, and being a protected area. These comparisons or even faint associations are partly modern, but partly they were already made in the Middle Ages; other places have been connected to paradise, and even in the Middle Ages the biblical paradise could be called by different names. The biblical word paradise – which appears in the Vulgata as well as in the Septuaginta in 1 Moses 2:8 (paradisus/ παραδεισος) – was translated by Martin Luther and in the King James' translation, as "garden of Eden." One example (fig. 6), the so-called Velletri or Borgia world map (which clearly reveals great knowledge about antiquity[7]), represents the *locus deliciarum* – the Latin translation of the Hebrew "Eden" according to Isidor of Seville.[8] In this paper I am dealing with some parts of a kind of medieval semantic field of "paradise island," trying to bring together islands that have explicitly been identified or described by any kind of recognizable comparison as paradise islands. I will show several examples of this because all these island descriptions and traditions could be mixed at some point and any of the descriptions could raise expectations for any of the islands. More specifically, the features of terrestrial paradise could be connected mentally to any paradisiacal island encountered anywhere in the world. At the same time, in this brief contribution to a specific conference topic it is not possible to trace and reconstruct all sorts of lines medieval authors followed or may have followed when writing about any particular island connected in some way to ideas of paradise islands. What I am trying to do is to provide insight into which ways could have been followed, which mental associations could have been connected, which identifications and explanations where considered[9] – and should be considered by scholars dealing with geographical interpretations of "paradise."

---

7 On this map and its special features concerning historical and eschatological considerations see Felicitas Schmieder, "Christliche Weltherrschaft – Anspruch und Grenzen in den Raum gezeichnet. Die Velletri/ Borgia-Karte (15. Jh.) in ihrem ideengeschichtlichen und politischen Kontext," in *Herrschaft verorten*, ed. Baumgärtner and Stercken.

8 Isidor, *Etymologies* XIV, 3, 2 = t. II, p.112.

9 I am also not interested here in placing geographical ideas of paradise into the whole range of medieval images of paradise, not the least the spiritual metaphysical ones. For this, see, e.g., Julia Eva Wannenmacher, "Geographie des Unendlichen. Räumliche Vorstellungen von Paradies und Infernum im Mittelalter," in: *Frömmigkeit – Theologie – Frömmigkeitstheologie. Contributions to European Church History. Festschrift für Berndt Hamm zum 60. Geburtstag*, ed. Gudrun Litz, Heidrun Munzert, and Roland Liebenberg (Leiden: Brill, 2005), 47-63; Reinhold R. Grimm, *Paradisus Celestis, Paradisus Terrestris* (Munich: Fink, 1977).

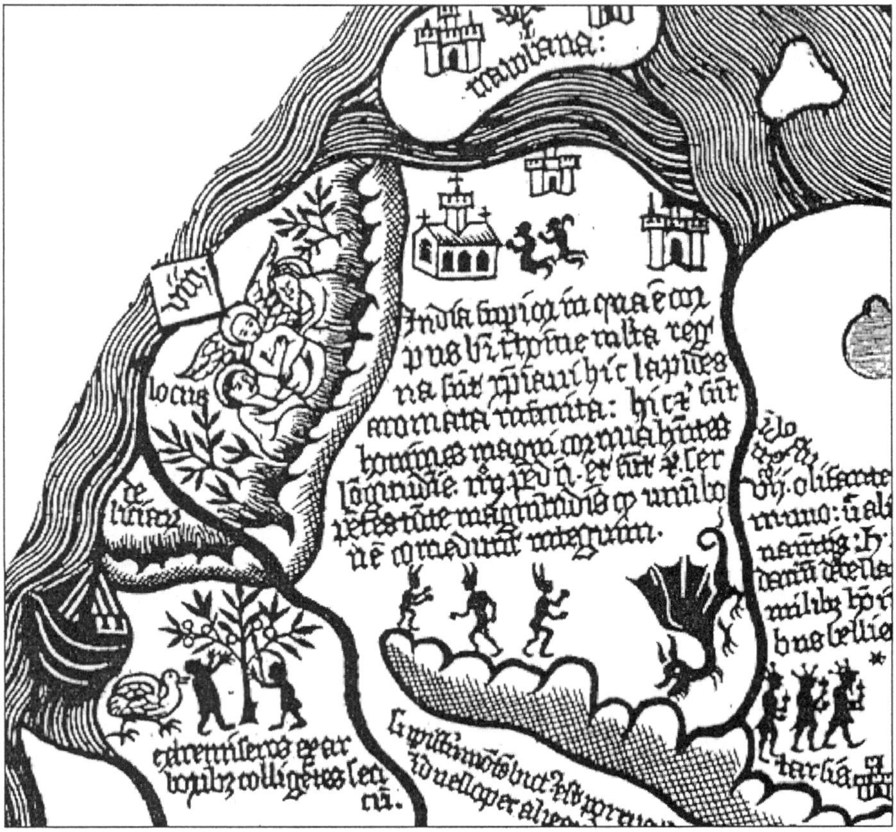

Fig. 6: Paradise as *locus deliciarum* on the Velletri/Borgia map, Vatican museum Borgiano XVI (Ø 6,4 m) 15th c. (from Robert Almagià, ed., *Planisferi, carte nautiche e affini dal secolo XIV al XVIII esistenti nella Biblioteca Apostolica Vaticana*, Città del Vaticano: Biblioteca Apostolica Vaticana, 1944, 27).

The best-known islands that have been explicitly compared to, if not identified with, paradise are the *insulae fortunatae* (or *fortunarum, fortunatarum*). Comparison and identification could happen although these islands were known to be in the Western ocean, where people looked for them and actually, in contemporary perception, found them at the beginning of the fourteenth century, when the Canary Islands were reached and taken by Latin Europeans. At the end of the fourteenth century, Domenico Silvestri described:

> … they carry nearly everything good, so to say lucky and blessed by a copiousness of fruits. By their nature fruits are brought about from precious forests, the hills are clothed with grapes no-one has planted, a great variety of herbs, grain and vegetables grow for the people. Therefore the error of the heathens

and the songs of the secular poets due to the fecundity of the land believed they were Paradise (*easdem esse Paradisum putaverunt*).[10]

This is, almost literally, a quotation from Isidor of Seville[11] on these islands (although the reference given names the twelfth-century geographer Guido of Ravenna). May Isidor have noted it all as a heathen error? By repeating it time and again medieval authors also repeated the convincing features that had created the error in the first place, and they repeatedly mentioned the *insule fortunate* and paradise as suggestively close to each other. And this may have had consequences, as we will be discussed below.[12]

The just-quoted late medieval author, Domenico Silvestri from Prato near Florence (writing around 1385-1406),[13] put together a huge collection, *De insulis et earum proprietatibus* (On islands and their characteristic features), in order to complement Giovanni Boccaccio's *De montibus, lacubus, fluminibus, stagnis et paludibus et de nominibus maris*[14]

---

10 Domenico Silvestri, *De insulis et earum proprietatibus*, ed. Carmela Pecoraro, *Atti della Accademia di scienze, lettere e arti di. Palermo* 14.2 (1954): 117-8: *Fortunate insule vocabulo suo Guidone Ravennate scribente significant omnia fe[r]re bona, quasi felices et beate fructuum ubertate. Sua [enim aptae] natura pretiosarum poma silvarum parturiunt, fortuitis vitibus iuga collium vestiuntur, ad herbarum vices messis et [h]olus vulgo est. Unde gentilium error et secularium carmina poetarum propter soli fecunditatem easdem esse Paradisum putaverunt. De his fortasse sensit Virgilius in sexto inquit: Devenere locos letos et amena virecta Fortunatorum nemorum sedesque beatas. Site sunt autem in oceano contra levam Mauritanie, occiduo proxime et inter se disiecto mari discrete. Quorum ut refert Solinus, sex nominibus narrantur: Embriona scilicet, nullis edifitiorum vestigiis, ut Plinius scribit, et huius in montibus esse stagnum cum arboribus similibus ferule ex quibus aqua exprimitur, e nigris amara, ex candidioribus potui iucunda. Due alie, quibus Iunionie nomen constat, licet Martialis tertia vocat Ceodem, quarta Capraria, Nivaria quinta, sexta Canaria. In aliqua istarum arbores crescere prohibent usque ad CXL pedum altitudinem, ibi avium multitudo, nemora pomifera cariotas ferentia, ibi copia larga mellis, et lactis alieque res abundant prout in unaquaque tangitur. Hec secundum alios Occidentales dicuntur. Ad has ut Petrarca recitat nuper armatum Ianuensium penetravit navigium et Clemens sextus illi patrie principem primum dedit, quem, vidisse Petrarca testatur Hispanorum et Gallorum regum misto sanguine generosum quendam virum. Gentem autem harum insularum refert pre cunctis ferme mortalibus gaudere solitudine, tamen incultam moribus et [ideo] belluis similem et nature magis instinctu quam electione aliqua [vi]vere et in solitudine errare cum feris seu suis gregibus.* (the text taken from Isidor is undelined; see n. 11).

11 Isidor, *Etymologies* XIV, 6, 8; t. II, p.133: *Fortunatarum insulae vocabulo suo significant omnia ferre bona, quasi felices et beatae fructuum ubertate. Sua enim aptae natura pretiosarum poma silvarum parturiunt; fortuitis vitibus iuga collium vestiuntur; ad herbarum vicem messis et holus vulgo est. Unde gentilium error et saecularium carmina poetarum propter soli fecunditatem easdem esse Paradisum putaverunt. Sitae sunt in Oceano contra laevam Mauretaniae, occiduo proximae, et inter se interiecto mari discretae.* – The differences (see n.10) may be due to the fact that Domenico did not take the piece directly from Isidor. Among other things, Domenico or his direct source complemented the description – not the least by interpreting one *poeta saecularis* – by connecting it to recent events and geographical exploration.

12 Scafi, *Mapping*, 145 speaks about Isidor's "statement …that the Garden of Eden was not to be confused with the classical Fortunate Islands," but it is much less than that, much more an implicit verdict than an explicit warning not to confuse them.

13 For him, see Marcia Milanesi, "Il *De insulis et earum proprietatibus* di Domenico Silvestri (1385–1406)," *Geographia antiqua. Rivista di geografia storica del mondo antico e di storia della geografia* 2 (1993): 133-46.

14 Giovanni Boccaccio, *De montibus, silvis, fontibus, lacubus, fluminibus, stagnis seu paludibus, de diversis nominibus maris*, ed. Manlio Pastore-Stocchi, in *Tutte le opere*, vol. 7/8, 2, ed. Vittorio Branca (Milan: Mondadori, 1998), 1815-2122.

(On mountains, lakes, rivers, ponds, and swamps and on the names of the seas) – obviously leaving out the islands – which had been written (around 1355-1360) as a dictionary of geographical places handy for commenting on Classical Latin poets especially. Alphabetically and analytically, Domenico deals with all the islands he knows on the most up-to-date level of geographical knowledge, which was based on antique information as well as recent discovery. Under the letter "I" he not only enumerates the *insule fortunate*, but also each of the six of them under its own name.[15]

One of them shows how Domenico worked – and this makes him a typical witness for late medieval dealings with far-away islands in the Okeanos that was believed to somehow surround the inhabited earth. Domenico uses *Canaria* to indicate two explicitly different islands: "The island of *Canaria* is called after the dogs which live there in great number. It is situated in the oriental [sic] ocean and is one of the *insule fortunate* about which I will say more below ..."[16] And then: "The island of *Canaria*, which is different from the former, lies beyond the columns of Hercules [normally the straits of Gibraltar] and is one of the found islands ..." – that is to say: found lately – since for both islands Domenico refers to recent naval expeditions which, in reality, all reached the same archipelago still called the Canaries.

The doubling of *Canaria* shows that the finding and clear identification of traditionally "known" geographical items did not, in the later Middle Ages, lead to abandoning different readings. Even after a successful, seemingly definite, identification there was room for alternatives, and different explanations could be placed side by side if harmonisation was not easily possible. The traditional material, authoritative in principle, was apparently preferably kept in storage. And this was even possible with islands that were, in Domenico's time and as he well knew, ruled by Latin Europeans and on whose identification people normally agreed. For example, the influential *Atlas Catalan* (by Abraham and Jafuda Cresques, 1375, may also have helped to create the model for the shape in which the islands were represented frequently from then on; fig. 7-9) located the *iles beneventurands* or *yles fortunades*, quoting Isidor, at exactly the place were the Canaries had been found and where they are known up to today.[17] Domenico himself

---

15 Domenico Silvestri, *De insulis*, 65 (*Canaria*), 66 (*Capraria*), 104 (*Embriona*), 140 (*Iunonia*; two islands with the same name in the very same arcipelago; while the alternative name for one of them, *Ceodem*, has no separate entry), 173 (*Nivaria*).

16 Domenico Silvestri, *De insulis*, 65 (*Canaria*).

17 J.A.C. Buchon and J. Tastu give an edition of the inscriptions in the *Atlas*, together with a French translation ("Notice d'un atlas en langue catalane manuscrit de l'an 1375, conservé parmi les manuscrits de la Bibliothèque Royale, sous le n° 6816, Fonds Ancien, in Folio Maximo," *Notices et extraits des manuscrits de la Bibliothèque Nationale* 14, No. 2 (1841) : 1-152, here 67-8). Although the transcription does not specify where abbreviations are solved, and could be discussed by different readings in detail, I quote it here: "Les Iles Beneventurades son en la mar gran contra la ma squera prop lo terme del Occident mes prop son dintre la mar. Isidori ho diu al seu XV libre que aquestes son dites Beneventuradas quar de tots bens blats fruyts herbes arbres son plenes e los pagans se cuiden que aqui sia parais per lo temprament del sol e

added, in his entry on the *insule fortunate* themselves (of which the beginning has been quoted above), that the *insule fortunate* can be found in the ocean across from the Mauritanian shore close to sunset (*Site sunt autem in occeano contra levam Mauretanie occiduo proxime* ...).[18] He may not have known the location of Mauretania too well and even more may have confused the oceans since he may not have been too secure in juggling directions – a solution which would fit quite well with the thinking of most late medieval Latin Europeans who did not travel themselves and were easily confused by geographical interrelations.[19]

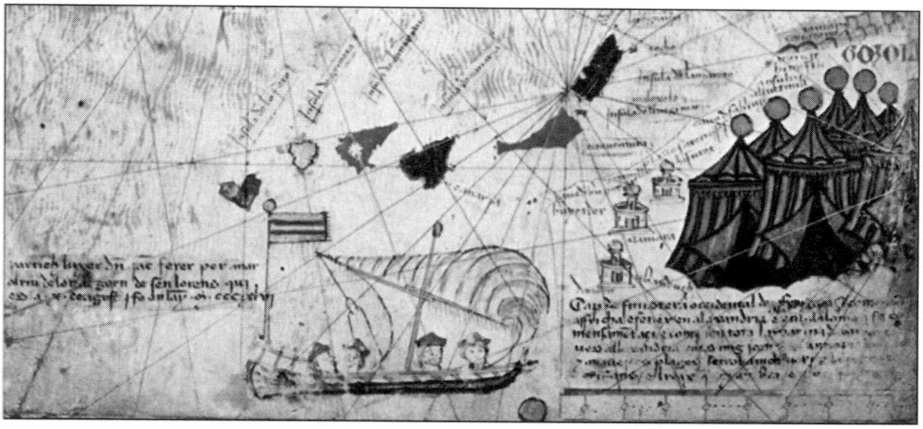

Fig. 7: Atlas Catalan (6 plates, each 69 x 49 cm), 1375 (from *Atlas Catalan von 1375*, ed. and transl. Hans-Christian Freiesleben, cf. n. 3, plate 2): the Canary islands

habundancia de la terra. Item diu Isidorius que los arbres hi crexen tots al meyns cxl pes ab molts poms e mols aucels. Aqui ha mel et let majorment en la ylla de Capria que ayxi es apellada per la multitud de les cabres que hi son. Item es apres Canari ailla dita Canaria per la multitut dels cans que son en elha molts grans e forts. Diu Plinus maestre de mapa mundi que en les Iles Fortunades ha un ailla un se leven tots los bens del mon com sense semrar e sens plantar leva tots fruyts. En les altees dels monts los arbres no son nulh temps meyns de fulla e de fruyts ab molt gran odor. Dasso menyen una part de lany puis segen les messes en loch dherba. Per aquesta raho tenen los pagans de les Indies que les lurs animas con son morts sen van en aquelles iles e vieun per tots temps de la odor daquels fruyts e allo creen que es lur paradis. Mes segon veritat faula es." Cf. the facsimile *Atlas Catalan von 1375,* ed. Freiesleben, plate 3. – The exact shape of the Canaries seems to have been shown earlier on the so-called Laurentiana-Medicean World Atlas, which usually is dated to 1351: Nordenskiöld, Periplus, cf. n.3, p.115.

18 Domenico Silvestri, *De insulis*, 117-8.

19 The multiplication of possibilities certainly did not make life easier; the famous Venetian cartographer Fra Mauro, on his world map from 1448-59, noted on the Canaries: *queste insule secondo alguni se dicono fortunate* [Piero Falchetta (ed.), *Fra Mauro's World Map* (Turnhout: Brepols, 2006)], no. 1238 – and then a second time close to Ireland: *queste insule de Hibernia son dite fortunate* (no. 2223).

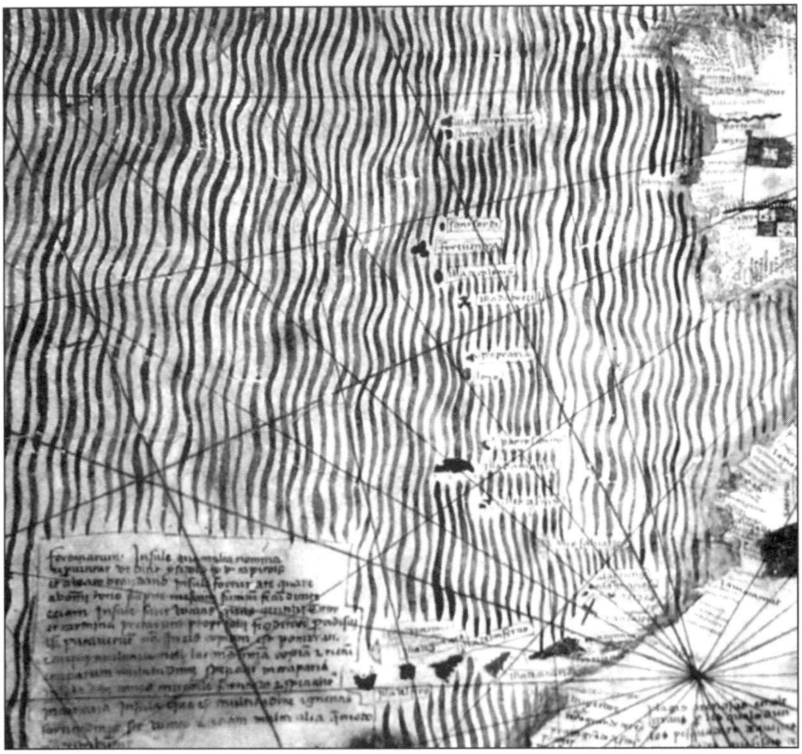

Fig. 8: Canary islands (*fortunatum insule*) and *illa de brezill* (formed like a cross, sixth in the row from top down) on the Catalan Estense world map at Modena (Ø 1,13 m), c. 1450 (from *Mappamondo Catalano Estense*, cf. n. 3).

This impression is strengthened by a further speculation by Domenico on *Canaria* which connects it with another crucial field of my topic:

Perdita is an island in the Indian Ocean and it is distinguished above all others by its mellifluence and its fecundity in all things. It is unknown to man, if it is not identical with *Canaria*, about which I talked above, which in our time was found – but it is said about *Perdita* that it was once found, later searched for and not found again, and this is why it is called 'lost island.'[20]

---

20 Domenico Silvestri, *De insulis*, 183: *Perdita insula Indico oceano sita amenitate omniumque rerum fertilitate pre cunctis longe prestantissima. Hominibus incognita est nisi esset Canaria, de qu supra, que nostro tempore fuit reperta, licet etiam dicatur ista aliquando inventa, postea quesita non est reperta. ideo dicitur Perdita. Ad hanc venisse Brandanum asserit Isidorus De ymagine mundi.* This was, actually, not told by Isidor, but by Honorius Augustodunensis in HIS De imagine mundi libri tres, book I, ch. 36, ed. Migne Patrologia Latina 172, col. 133A.

Thus, there may be three locations of two different *Canaria*... One can well see here how Domenico struggles with traditions and recent knowledge in a way typical for his collection.[21] He does not state why he tries to identify *Perdita* with *Canaria*, but it may have been for the copious abundance also found there – the feature which had prompted, as quoted in Domenico's opinion, some people to consider it all Paradise. This may be the context in which he adds: "*Perdita* was reached by *Brandanus* as is told by Isidore."[22]

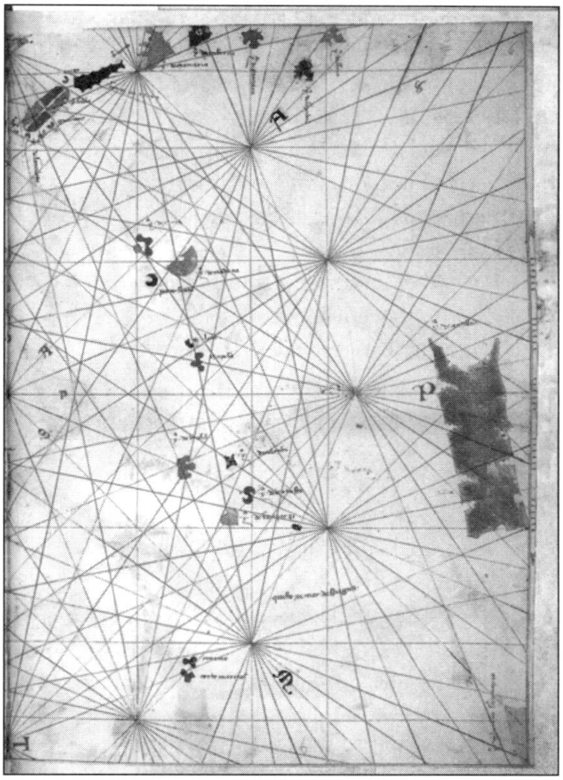

Fig. 9: Andrea Bianco, *Atlante nautico* (10 double folia, each 25 x 36 cm), 1436, tavola 5, right half (from Andrea Bianco, *Atlante nautico*, cf. n. 3): It shows part of the Atlantic off the Western coasts of Spain and North Africa, with the Canaries (on top of the map, which is oriented to the South), *ysola de brasil* (in shape of an "E") and *ysola de antillia* (a big red island on the very right side of the sheet).

21 On this struggle see Felicitas Schmieder, "'Den Alten den Glauben zu entziehen, wage ich nicht ...' Spätmittelalterliche Welterkenntnis zwischen Tradition und Augenschein," in *Autorität und Wahrheit. Kirchliche Vorstellungen, Normen und Verfahren (XIII. – XV. Jahrhundert)*, ed. Gian Luca Potestà (Munich, forthcoming). In the case of the Canaries and Perdita he takes part in a contemporary discussion (without explicitly interfering): Milanesi, "Il *De insulis et earum proprietatibus*," 140.

22 See note 20.

Here again are the suggestive paradisiacal features of the *insule fortunate*; the repeated quotation of Isidor's description kept the association alive. And additions like the one by the *Atlas Catalan* that "Indian heathens" would also believe the islands their paradise were not at all weakened by the assertion that this was nothing but a fabulous story. In this context, it was easy to insert Brandanus (or Brendan).

With this name, Domenico is alluding to a legendary tradition of not-entirely-clear origin (certainly not told by Isidor), but with strong Irish connections – and with huge impact on medieval ideas of the possibility of finding a paradise island in the Western ocean (which shows, again, that Domenico seems to mix up the oceans and directions). Starting with the *Navigatio Sancti Brendani*, a text from the tenth century at the latest, news spread throughout Latin Europe about the Irish monk, Brendan. He reportedly travelled by ship, together with 17 companions, from Ireland to the west, searching for a Paradise island, the promised land of the saints (*terra repromissionis sanctorum*), on their way to the promised land, the land the Lord has prepared for his saints from the beginning of the world as a place of refuge when the great persecutions finally overtake Christendom.[23] On their way, which took seven years, they came across several islands (which normally do not have names, but typical qualities) and found all sorts of wonders and monstrosities; the text combines Irish motifs of peregrination and apocalyptic expectations with presumably real experiences.

However real the memories of Irish sea travels to the West may have been – scholars have even considered an arrival in America, Brendan's islands found a place on several world maps among the islands that were represented on them all around the inhabited earth. Since he was reaching paradise, and the *insule fortunate* were closely connected to paradise islands, he ends up having reached the former. If, for Domenico Silvestri, the connection included many "maybes," this was not true for other examples. The famous Hereford Map, for example, created in England around 1300 and therefore especially familiar with traditions in its own region, explicitly and laconically identifies the *fortunate insule* with the Islands of Saint Brendan.[24] The fifteenth-century Modena map draws the connection at more length by mentioning all the old information about the *insule fortunarum/fortunate*: "The *fortunarum insule* are found with many names as says Isidor...and are called by Saint Brandan *insule fortunate* because ... [and then follow the

---

23 For the various versions and vast literature on the legends of St. Brendan, see Glyn S. Burgess and Clara Strijbosch (ed.), *The Brendan Legend. Texts and Versions* (Leiden: Brill, 2006); Clara Strijbosch, *The Seafaring Saint. Sources and Analogues of the Twelfth-Century Voyage of Saint Brendan* (Dublin: Four Courts Press, 2000).

24 *Fortunate insulee sex sunt insule Sancti Brandani*, in: Westrem, *The Hereford Map*, 389, nr. 987. Instead of six, the northernmost of the islands off the African coast is denominated this way (image section 10 in Westrem). – A similar identification can be found on the Pizigano map, and is discussed by Mario Longhena, La carta dei Pizigano del 1367 (posseduta della Biblioteca Palatina di Parma), in: Archivio storico per le provincie Parmesi ser. 4, 5 (1953) 25-130, here 58-61.

usual paradisiacal features]."[25] Even Fra Mauro, *Cosmographus Incomparabilis* from Venice, includes Brandan's *insula inferni* (without mentioning the traveller's name) at approximately the same time as the seventh islands among the Canaries – and, it may be said – thus close to paradisiacal islands.[26]

All these different manifestations are connected to circumstances already visible in Domenico Silvestri's compendium: islands were reached, lost, and found again. Because certain islands were included in written and painted collections, people had them in mind when going on naval expeditions, the texts and maps made people remember any promised or promising island, made them reflect on them, compare them, identify them, even try to find them. Among the many islands that entered the medieval imaginative collection that were expected to be found when going westwards and that could be connected, among other things, with paradisiacal abundance and other features, some were actually found around 1500 – that is, islands reached by seafarers borrowed (for one reason or another that cannot be traced in detail here) their new names from them. Not only the Canaries witness that; on several late medieval maps (among them the *Atlas Catalan* and the Modena map) we find one, two or even three islands called *brezill, berzil, barçil* or *brazi(l)*, normally off Ireland to the west and off the Portuguese western shores (and maybe also connected to Saint Brendan).[27] The Venetian Andrea Bianco, who also had good contacts with the Iberian peninsula, differentiated these two islands in his atlas; while *berzil* remains close to Ireland, off Portugal we now find *brasil* (thus the version with a future) on the way from the Iberian peninsula

25 *Il Mappamondo Catalano Estense*, ed. Milano and Battini, 147: *Fortunarum insule que multa nomina rep(er)iuntur ut dicit ysodole .lxv.* [*l(ibro) xv*] *capitols et abeato* [*a beato*] *brandano Jnsule fortunate quare ab om(n)i bono p(re)ciprie mesium furasu(m) fecu(n)ditate eciam jnsule sunt vocatis quas gentili* [*gentili(um)*] *Ettor* [Error] *et carmjna pretarum* [*sic!* on the map instead of *poetarum*] *propt(er) folij fe(cun)ditate p(ar)adisum ese putaveru(n)t na(m) in eis copiam est pomorum eaujum multitudo mell lac ma*[-]*sima(m) copia(m) (et) rietu(m)* [*rictum* ?] *eraparum* [*craparum*] *multitudinem spezialit(er) in craparia jnsula ubi canes mirabile fortitudo (et) specialit(er) in canaria insula qua est multitudine ignentis fortitudinjs sic vocata et eciam multa alia q(uae) modo no(n) describu(n)tur*" (my modifications of the transciption in [ ]). It is the only Latin text on the map (everything else is Catalan), so may be taken directly from some Latin source.

26 1448/59: He has *lanziloto, forteventura, gran canaria – isola de linferno – isola de fero* [Hierro], *la gomera und palma*, see Falchetta (ed.), *Fra Mauro's World Map*, no.1256, 1250, 1251, 1248, 1242, 1252, 1254.

27 Without having checked all the surviving maps, one can list *berzil* and *barzil* for the *Atlas Catalan* (1375), and *berzill* twice for the Modena map (c. 1450), also on the Laurentiana-Medicea World Map, quoted in note17. The presumably earliest and at the same time most copious version seems to be found on the portolan by Domenico Pizigano from 1367: Longhena (cf. note 24), 61; cf. also Arthur Dürst, "Brasilien im frühen Kartenbild," in *Brasilien – Entdeckung und Selbstentdeckung*, (Bern: Benteli, 1992), 38-48, who gives re-drawings of two maps without quoting any useful image of the original (on p. 38). – The name may have been the result of an adaptation of Saint Brendan's "fire island" by speakers of a Roman language to their expectations in the West, since fire or better blaze means, in Portuguese/ Spanish, *brasa* (Italian *brace*/ *bragia*).

towards America[28] (fig. 8-11). A possibly early medieval Iberian legend speaks of the island of *Antilia* (represented as a huge island at the western rim, again by Andrea Bianco, fig. 9), a place of refuge for seven Spanish bishops fleeing from the Muslim invasion (a scenario comparable to the reason which drove Brendan out on his search). The acceptance of names was frequently connected to the paradisiacal hopes of the discoverers (although this cannot be traced here in more detail).

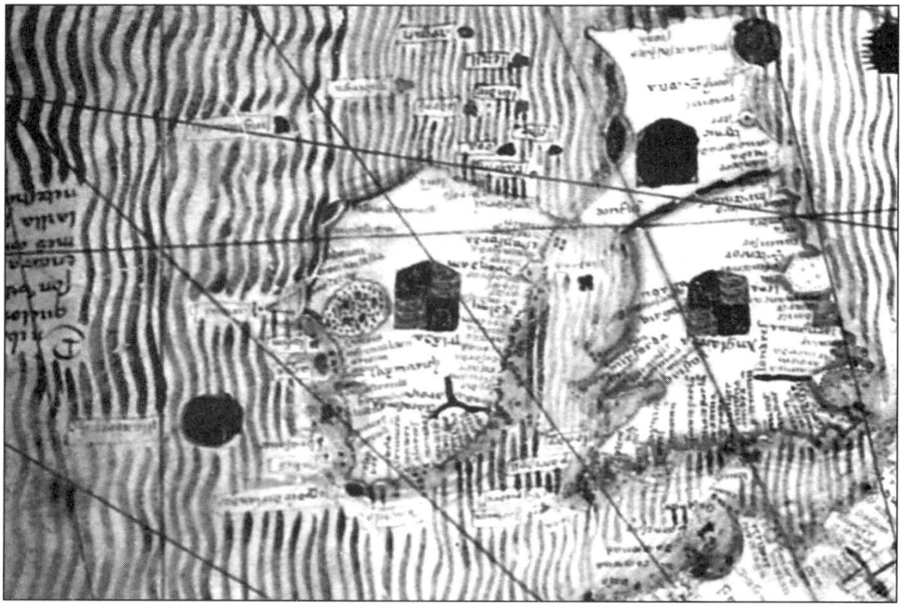

Fig. 10: Island of *brezill* on the Catalan Estense world map at Modena (Ø 1,13 m), c. 1450 (from *Mappamondo Catalano Estense*, cf. n. 2).

There are reasons for looking for these islands, not the least paradise, which go beyond geographical curiosity and correctness. Medieval distant travel rarely or never had only one motivation; economic reasons were added to military ones and pure curiosity – and to spiritual hopes. Especially the idea of terrestrial paradise hinted at more than a geographical, zoological or ethnographic meaning, to more than just riches, as interesting as these may have been at any time – paradise was not just one more paradisiacal place on earth. The journey of Brendan in the early Latin version may have had origins in real memories, but besides that it had eschatological features. Moreover, the *Navigatio* could easily, in the Anglo-Norman twelfth century, be translated or, better,

[28] Andrea Bianco was a collaborator of the famous Fra Mauro, who worked for the Portuguese court, at about the same time as the makers of the Modena map, cf. Edson, *World Maps*, 141-2 On Fra Mauro's map (1448-59), on the other hand, the only remaining island of *brezil* lies west of Ireland.

transformed into a pilgrimage of the soul seeking perfection. A little earlier, the German version had described travelling as a punishment for lack of belief – thus adding a moral sense to the literal one, which, of the four *sensus scripturae* well-known by educated medieval readers, we modern people have a strong tendency to rely on and reduce to. Paradise could be a geographical place and a spiritual place at the same time – a connection that was not at all strange in medieval thinking. On their way to find all the wonderful islands of different traditions people took with them ALL the meanings connected to the traditional names and to the connected ideas. Since, as has been noted above, islands and ideas were mixed, one concept of Paradise Island can rarely be separated from another. Brendan's paradise or the Garden of Eden, whether in the East or in the West, all became the same thing in the hopeful minds of travellers.

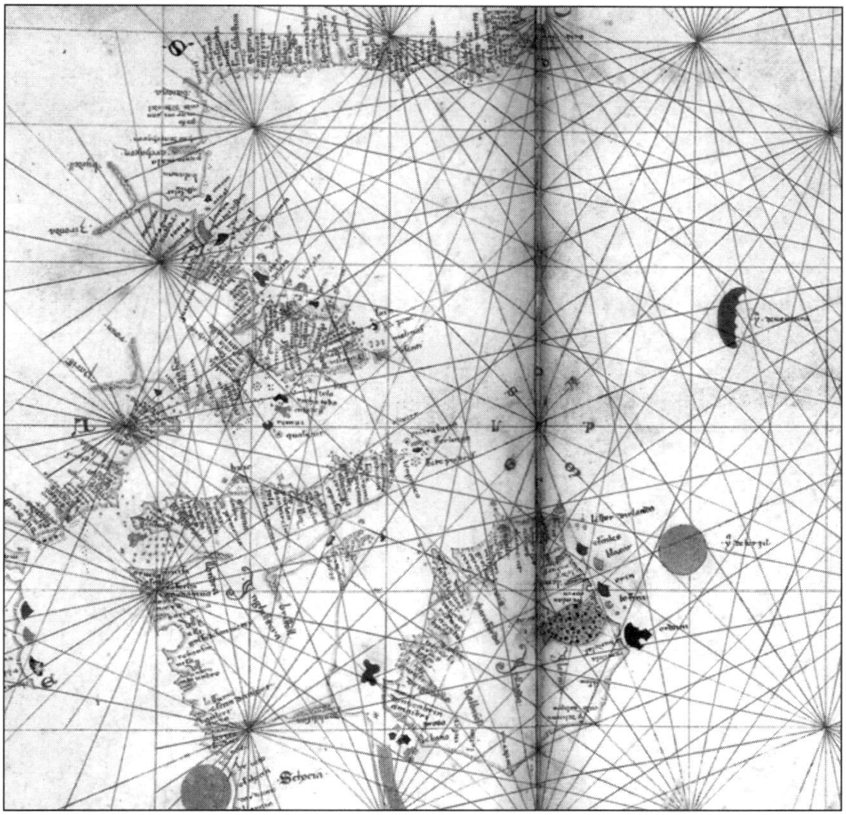

Fig. 11: Andrea Bianco, *Atlante nautico* (10 double folia, each 25 x 36 cm), 1436, tavola 6 (quoted from Andrea Bianco, *Atlante nautico*, cf. n. 3): Coast of France, Great Britain, Ireland and Northern Atlantic islands, among which a huge golden circle west of Ireland (map is oriented to the South) is named *ysola de berzil.*

Not least, the apocalyptic meaning of paradise was always present for educated medieval readers who had learnt to read the four senses of scripture into everything – in an allegorical-typological as well as an anagogical sense – suggesting that everything in the world was meaningfully pointing to something else, especially things past and present, and to future things. This eschatological geography underlying the *mappae mundi* was sometimes made very explicit on late medieval world maps; Andreas Walsperger makes the connection clear when he pictures the heavenly Jerusalem (that would come down to earth at the end of all times) as rising in terrestrial paradise (fig. 12).[29]

Thus, these maps were not made for pragmatic use in a sense of travelling by them, but they were made to show interpretations of history, present and future, political aims and spiritual hopes, all of which cannot be separated from each other as modern interpreters would like them to be. This was so closely linked that a "spiritual" registering of the world could be symbolically represented as *mappa mundi*, noting which places and people had become Christian and which were still awaiting it, thus registering the world eschatologically on its way to full conversion as an important precondition for the Second Coming of Christ. The fifteenth-century book *Mappemonde spirituelle* by Jean Germain is depicted, in the dedication miniature, as a world map. This inventory of the religious beliefs of the peoples of the world, which probably was never connected to an actual map, was written in the context of the Turkic crusade plans of Duke Philip of Burgundy (written by the first chancellor of the order of the *toison d'or*, also founded in this context). And Philip, as recipient of the dedication, points his finger exactly at Paradise on Earth, which is, hence, underlined as the spiritual-eschatological-geographical place on earth to which all the planning and achievement should lead (fig. 13).[30]

---

29 The first preserved globe from 1492/93 (thus on the very eve of Columbus' discoveries) by Martin Behaim, who was close to the Portuguese knowledge and knows the Azores, nevertheless keeps *antilia* (south and a bit west from the Azores) and an *insula de sant brandan* further to the South-West. In both cases only the contours of the islands are drawn, probably in order to notified that they were not yet found. The islands are accompanied by long texts describing the concerning legend. Far to the West of this all, close to the Chinese border, there is *Cipangu* (Japan as known by Marco Polo), which is (maybe not by chance) close in form to the *antillia* of Andrea Bianco: Focus Behaim Globus, 2 vols., Nurenberg 1992, best visible on one remake p. 757).

30 The image is taken from Patrick Gautier Dalché, "Zeit und Raum," in Jacques Dalarun (ed.), *Das leuchtende Mittelalter* (Darmstadt: Wissenschaftliche Buchgesellschaft, 2005), 21-51, here 45 [French original *Le Moyen Âge en lumière* (Paris: Fayard, 2002), whom I especially owe the hint at paradise-pointing. On this work and its position between traditional world view and "modern" geographical experiences, see David J. Wrisley, "Situating Islamdom in Jean Germain's *Mappemoden spirituelle* (1449)," *Medieval Encounters* 13 (2007): 326-46; Margriet Hoogfliet, "The Medieval Texts of the 1486 Ptolemy Edition by Johann Reger of Ulm," *Imago Mundi* 54 (2002): 7-18; François Berriot, "Images de l'Islam dans le Débat manuscrit de Jean Germain (1450)," *Bulletin de l'Association d'étude sur l'humanisme, la réforme et la renaissance* 14 (1981): 32-41.

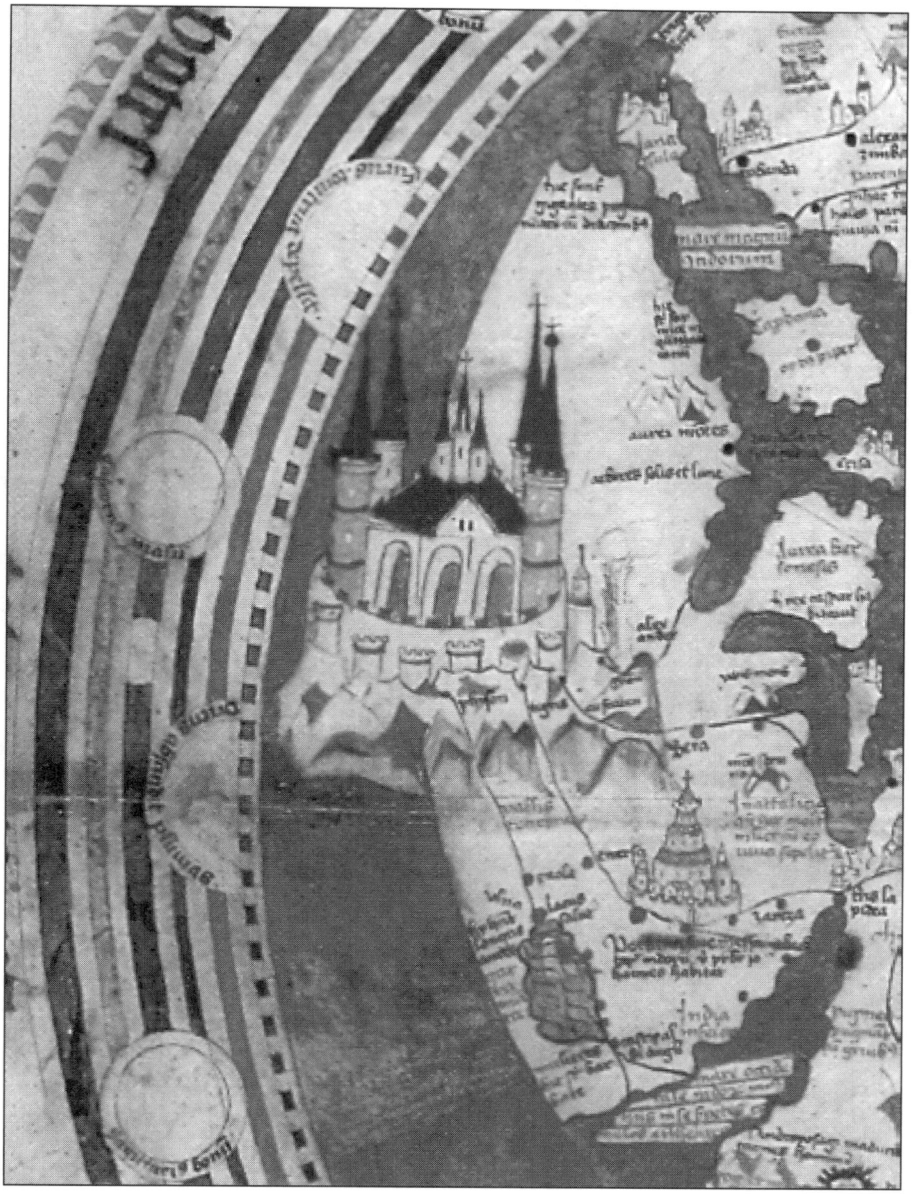

Fig. 12: Andreas Walsperger, World map (circle on 57,7 x 75), 1448, (cf. n. 3): Heavenly Jerusalem coming down to Earth at the End of Times into Paradise on Earth

Fig. 13: Detail from the dedication page of Jean Germain, *Mappemonde spirituelle*, 1449 (Lyon BM, Ms. Palais des Arts 32, fol. 1r; from Dalarun, Das leuchtende Mittelalter, 45)

Searching for the presumably well-known islands (in the sense of knowing what to expect of them), looking for fecundity and isolation, for blessedness and protective seclusion – always considering it possible that one of them might turn out to be paradise on earth was not the least motivation to make people travel, longing for riches and fertile land, for a place of refuge in dangerous times, but also longing for eternity and the place of God's peace. Thus, it may be provocative but certainly not wrong, when I conclude: When Christopher Columbus set sail to the Western ocean to reach the rich Indian islands he had in his mind images of paradise islands wherever they were located exactly, and he was, among other things, trying to find the eschatologically promising Paradise on Earth.

THE POWERS AND PURPOSES OF AN INSULAR SETTING – ON SOME MOTIFS IN OLD-NORSE LITERATURE

Kristel Zilmer

Introduction

The narrative representation of islands in Old Norse literature was shaped by various literary and cultural layers and born out of practical and symbolic, but also deeper, cognitive purposes. It has been argued previously that such imagery can highlight traditional Nordic cultural experiences, while at the same time conducting a dialogue with a medieval European understanding on the meaning of islands.[1] The analysis of examples derived from Old Norse prose and poetry has shown that, on the one hand, a pragmatic rationale exists behind certain island-related scenes in terms of mapping strategic sites along the characters' travel routes or within the main arena of action. On the other hand, one also encounters motifs that speak of the symbolic peculiarity of the island setting – a concept that is activated when islands are depicted as sites for extraordinary events and happenings.

In this paper I shall take a closer look at the latter facet of island representation – the emphasis will be laid upon depictions that connect an island setting with certain transformative (religious and magic) experiences which may affect a character in a decisive manner and, as such, carry deep significance for the portrayal of his further actions.[2] Through this analysis I will also approach the broader concept of an island as

---

1 See Kristel Zilmer, "Scenes of Island Encounters in Icelandic Sagas: Reflections of Cultural Memory," *Viking and Medieval Scandinavia* 4 (2008): 227-48; eadem, "On Some Pragmatic and Symbolic Features of Island Representation in Old Norse Literature," in *Approaching the Viking Age*, ed. Ērika Sausverde and Ieva Steponavičiūtė (Scandinavistica Vilnensis 2) (Vilnius: Vilnius University Press, 2009), 197-215. Whereas the first article concentrates on saga material, the second also brings in perspectives from skaldic and eddic poetry. Concerning the motif of fictional islands, see Else Mundal's contribution in this volume.

2 Thus, it is the more metaphorical dimension of island representation that is the point of depature here, which does not mean that other motifs would not be equally important. Indeed, as will be shown later, a

an entity that is characterized by both isolation and interaction and in this manner provides a perfect arena for processes that bring about change.

An introductory remark to make concerns the definition of an "island" in this study. Such a definition naturally has to depend upon how the sources themselves designate and depict islands, but at the same time it is necessary to take into account various contextual interpretations.[3] The applied terms and naming practices can provide us with suitable guidelines; in the meantime, the task is complicated by the fact that distinctions between islands and (main)lands cannot be drawn automatically either from the linguistic or from the contextual point of view. In fact, there is an inbuilt fusion of the two in the very word "island," in Old English *īgland*/*īegland*, since it combines the components "island" (i.e. "a thing in water") and "land."[4] With regard to the Old Norse sources, besides the word *ey*, we find, for example, *eyland*, similarly standing for "island-land."[5]

To illustrate further uncertainties in connection with experiencing a particular territory as an island or a (main)land and communicating such experiences through names, we can take a look at the case of Iceland, already in Old Norse labelled *Ísland* ("ice" + "land"). Thus, this name determines Iceland as a "land." However, it is of interest to mention that one of the supposed early names of Iceland contained the element signifying an island; according to *Landnámabók* (The Book of Settlements), the

---

certain fusion of symbolic and pragmatic perspectives is inevitable. For a discussion of islands as significant landmarks and navigational aids, see, for example, the articles mentioned above.

3 See, e. g., Zilmer, "Scenes of Island Encounters in Icelandic Sagas," 231-33; cf. also the semantic analysis of the Old Norse word *ey* (i.e., "island") as well as the overview of different collocations as provided by Johan Fritzner, *Ordbog over det gamle norske sprog* I, 2nd ed. (Oslo: Universitetsforlaget, 1973), 354; Richard Cleasby and Gudbrand Vigfusson, *An Icelandic-English Dictionary*, 2nd ed. (Oxford: Clarendon Press, 1957), 134.

4 See more about the etymological background of the Germanic word in Jan de Vries, *Altnordisches etymologisches Wörterbuch*, 2nd ed. (Leiden, Boston, Cologne: Brill, 2000), 106. Similar ambiguity is visible in modern island names in several languages; for example, in Swedish one finds the name Öland and in Estonian Saaremaa, both designating islands in the Baltic Sea, and meaning "island land."

5 An interesting example is found in the 13th-century Norwegian *Konungs skuggsjá* (The King's Mirror) where one wonders whether Greenland should be called an island (*eyland*) or a mainland (*meginland*). The explanation provided sees Greenland as a mainland due to the fact that animals live in Greenland that are not known from islands. See Rudolf Keyser et al, *Speculum Regale. Konungs-skuggsjá. Konge-speilet. Et philosophisk-didaktisk skrift, forfattet i Norge mod slutningen af det tolfte aarhundrede* (Christiania: Trykt hos C.C. Werner & Comp, 1848), 42. With regard to exemplifying territories that could be identified as islands in the Old Norse narrative tradition, a source to consult would also be the poetic lists (*þulur*) that even contain certain geographical names, including those of various islands. Such lists form an addition (or interpolation) to *Snorra-Edda* (The Younger Edda, also called The Prose Edda). Whereas *Snorra-Edda* dates back to the beginning of the 13th century, the exact age of *þulur* is, however, debated; see, e.g., Eyvind Fjeld Halvorsen, "Þulur," in *Kulturhistorisk leksikon for nordisk middelalder fra vikingtid til reformasjonstid*, ed. A. Bugge et al (Oslo: Gyldendal, 1976), 404-5.

name given to the country by the Swedish traveller Garðarr Svávarsson was *Garðarshólmr* (that is, the Isle/Holm of Garðarr).[6]

In this current context, I shall not engage in a further semantic and/or etymological discussion of naming practices in connection with islands, but rather operate with a more general understanding of the concept of *ey*. A point made in the manuscript *Codex Upsaliensis*, DG 11 of *Snorra Edda* serves as a suitable premise: "… *en ey heitir þat land sem sior eða vatn fellr vm hverfis*."[7] The main idea thus underlines the experience of an island as a piece of land surrounded by water (which simultaneously accords with the etymological understanding referred to above).[8] Based upon such criteria, an island can be characterized as a clearly distinguished setting in itself – it is separated from other territories in terms of natural boundaries, but it can also enter into wider communicative networks. As one can notice from the motifs present in narratives, on the one hand, islands emerge as places where one may encounter/undertake something out of the ordinary, and they may even serve to create a certain distance between one person/group and others. On the other hand, islands are often attached to other, either close or more distant, territories, allowing for various contacts and influences.

Islands – symbolic and real – the ambiguous case of Prímsignd /Prímsigð

In order to illustrate some features of the island setting, I shall start with a short passage from one of the redactions of *Jómsvíkinga saga* (The Saga of the Jomsvikings): *Hjǫrungavágr er svá háttaðr at sker liggr í miðjum váginum en ey fyrir norðan er Prímsignd heitir; en Hǫrund liggr fyrir sunnan.*[9] This saga commentary – from an edition based upon the Stockholm 7 qv. manuscript version – includes references to a skerry,

---

6 See Jakob Benediktsson, *Íslendingabók. Landnámabók* (Reykjavík: Hið Íslenzka Fornritafélag, 1968), 34-35. About the naming of Iceland, see, e.g., also Judith Jesch, "Geography and Travel," in *A Companion to Old Norse-Icelandic Literature and Culture*, ed. Rory McTurk (Malden MA, 2005), 119-135.

7 Jón Sigurðsson, *Edda Snorra Sturlusonar II. Edda Snorronis Sturlæi II* (Hafniæ: Sumptibus Legati Arnamagnaeani 1852), 366. "… an island is a land which is surrounded by a sea or lake" (my translation).

8 At the same time, a deeper understanding of islands need not be limited to waterbound entities; indeed, in the Middle Ages one could also experience remote and somehow peculiar places as islands, cf. John Gillis, *Islands of the Mind. How the Human Imagination Created the Atlantic World* (New York,: Palgrave Macmillan, 2004), 17.

9 Norman F. Blake, *Jómsvíkinga saga. The Saga of the Jomsvikings* (London: Nelson, 1962), 32: "The layout of Hjǫrungavágr is such that a skerry is situated in the middle of the creek and there is an island called Prímsignd to the north of it. Hjørund lies to the south," On the redactions, see, e.g., Blake, "Introduction," in *Jómsvíkinga saga*, xv-xxi, and Ólafur Halldórsson, "Jómsvíkinga saga," in *Medieval Scandinavia: An Encyclopedia*, ed. Phillip Pulsiano et al. (New York, London: Garland, 1993), 343-344.

an island called *Prímsignd* (alternatively *Prímsigð*),[10] and another one called *Hǫrund*. Explicit information about *Hǫrund* being an island is recorded in AM 510 qv., which also mentions the fjord *Hǫrundarfjǫrðr* (most likely the modern Hjørundfjord in the region of Sunnmøre). The fjord is further referred to in AM 291 qv. and in *Flateyjarbók*, where the context similarly demonstrates that *Hǫrund* itself must be an island.[11] The landscape is outlined in a realistic manner – it seems to have been important for the saga writer(s) to depict a particular arena for action and focus upon what they considered to be its characteristic features. With regard to the general storyline in the saga, the question has naturally been where the battle of *Hjǫrungavágr* – which marks the climax of the whole story – might have taken place.[12] As part of such a search, scholars have attempted to locate the abovementioned islands. With regard to *Prímsignd/Prímsigð*, some suggestions have been the islands of Sula and Vigra off the west coast of Norway.[13] However, the saga's *Prímsignd/Prímsigð* did not have to refer to a specific and well-known island. The name might have been attributed to a place that served to make a point in the narrative.[14] That is to say, we may here witness a deliberate combination of landscape features with narrative constructions that mediated a particular message.

To us, the name in its recorded forms appears ambiguous. On the one hand, it may mean "Primsigned," i.e., "cross-marked," and as such be associated with the Christian practice of *prima signatio* or some other form of religious blessing or consecration.[15] On the other hand, it may have been inspired by the shape and features

---

10 In preserved manuscripts the name is recorded as *primsigþ, primsigd* (AM 291 qv. from the latter half of the 13th century; *Flateyjarbók* from the 14th century) or *primsignd* (Stockh. 7 qv. from the early 14th century; AM 510 qv. from the late 16th century); see Didrik Arup Seip, "Primsigd – et gammelt øynavn," *Maal og minne* 1959: 156-7. A Latin translation of the saga also exists from 1592-1593.

11 The name may possibly stand in connection with the similar-sounding fjord.

12 This approach follows the understanding that the saga reflects certain events that are to some extent based upon actual tradition. The present paper does not take a stand on the blend of facts and fiction in the saga, but see, e.g., Knut Helle, "Jomsvikingeslaget – islandsk heltediktning?" in *Kongsmenn og krossmenn: Festskrift til Grethe Authén Blom,* ed. Steinar Supphellen (Trondheim: Tapir, 1992), 167-193; Halldórsson, "Jómsvíkinga saga"; Alison Finlay, "History and Fantasy in *Jómsvíkinga saga,*" in *The Fantastic in Old Norse/Icelandic Literature. Sagas and the British Isles. Papers of the 13th International Saga Conference, Durham and York, 6th-12th August 2006,* eds. John McKinnell et al, volume 1 (Durham: Centre for Medieval and Renaissance Studies, Durham University, 2006), 248-57; Leszek P. Slupecki, "Facts and Fancy in Jómsvíkinga saga," in *The Fantastic in Old Norse/Icelandic Literature* 2, 906-15.

13 See, e.g., Seip, "Primsigd – et gammelt øynavn," 156-7; John Megaard, "Hvor sto 'Slaget i Hjǫrungavágr'? Jomsvikingeberetningens stedsnavn og Sæmundr fróði," *Alvíssmál* 9 (1999): 29-54, see particularly 33-36.

14 As an alternative to the larger and better known islands of Sula and Vigra, it has been suggested that the island in question may have been Eika, in the neighbourhood of Hareidlandet; see J. Ottesen, "Stod jomsvikingslaget ved Eiksund?" http://flyfotoarkivet.no/artiklar.php (2006, accessed 15. 03. 2010). According to Ottesen, Eika is a small island covered by forest, which would correspond to the description provided later in the saga.

15 See, e.g., Lee M. Hollander, *The Saga of the Jomsvikings* (Austin: University of Texas Press, 1955); Megaard, "Hvor sto 'Slaget i Hjǫrungavágr'?"; Ottesen, "Stod jomsvikingslaget ved Eiksund?". Megaard proposes

of some island (real or imaginary) that would remind one of either a cross or a sickle, to follow an alternative interpretation of the word as *Prímsigð*, that is, "the new moon's sickle."[16]

In this current context, I would argue that *Prímsignd/Prímsigð*, being an island, belongs within a specific scheme of events – its name, described position, and characteristic features have their own logic and motivation in the narrative. As stated in the introduction, islands can fulfill particular functions in the sources; the fact that they are included in depicting/creating a landscape is an important trait in itself. At the same time, I want to emphasize that the sources do not always find it necessary to identify every single site with a name; there is also mention of unnamed islands along the lines: "they took lee close to some island." When applying a name such as *Prímsignd/Prímsigð* to designate an island that forms part of the setting where an important battle is localized, there must have been certain motifs connected to that choice.

Many of the island references provided in the saga literature, i.e., in kings' sagas and the sagas of Icelanders but also in the skaldic poetry, fit into the scheme of events that are typical of the portrayal of Viking activities – islands emerge as strategic sites and central stopping places along one's travel and/or battle route.[17] A particular motif is the localization of various battles and campaigns within the island setting. Specific connotations of island fighting and duelling are, in fact, also provided through the very name for the act of duelling in Old Norse – the word *hólmganga* includes the component *hólmr*, which refers to small islets and holms.[18]

In the case of *Prímsignd/Prímsigð*, I would claim that the reference is important not only as an attempt to reconstruct a realistic landscape and in this manner outline the topography of a central battle arena, but also due to its symbolic role in the narrative, as indeed, reflected in its name. Examining the narrative logic behind the mention of *Prímsignd/Prímsigð* in the saga, we learn a little later that the key figure, Earl Hákon, went into a forest on that island, knelt down facing north and prayed so that he would have better luck in battle: *Þar kømr bœnarorðum hans at hann skorar á fulltrúa sinn, Þorgerði*

---

that the form *Prímsignd* (cf. Latin *prima signatio*) is a latinized version of Old Norse *vigðr* connecting with the island of Vigra (in Old Norse *Vigr*). He also discusses similar naming strategies as the background for *Hǫrund*, which he identifies with the island of Sula; the argument is based upon semantic similarities between the word *hǫrund* ('penis') and *súla* ('post, pole'); for a detailed discussion, see 33-41.

16 See, e.g., Seip, "Primsigd – et gammelt oynavn."

17 See more about this aspect of island representation and its cultural-historical relevance in Zilmer, "Scenes of Island Encounters in Icelandic Sagas," 233-7.

18 See Zilmer, "On Some Pragmatic and Symbolic Features of Island Representation." Some of the island duels even take place between a man and a supernatural being of some kind. A comment occurring in *Vǫlsunga saga* (The Saga of the Volsungs) brings in a deeper mythological dimension, claiming that the final battle site of Ragnarok will be an island called *Úskaptr*, "the Uncreated;" see Wilhelm Ranisch, *Die Vǫlsungasaga* (Berlin: Mayer & Müller, 1891), 30.

Hǫlgabrúði.[19] At first the goddess did not respond to his prayers and also refused the sacrifices that were offered to her. Finally, she accepted the sacrifice of Hákon's own son.

Thus, *Prímsignd/Prímsigð* gains a deeper meaning in the narrative due to the act of praying and the sacrificial ritual that is said to have taken place there. In view of this, even if the island as such is completely fictional, it would be understandable why a name expressing certain religious and/or cultic connections would have been applied. This might then speak in favour of interpreting the name as *Prímsignd*, the Primsigned. The possible Christian connotations attached to such a name may bear witness to the intentional creation of a certain tension and transformation process between the old and the new, which is also visible from the reflected scene. Earl Hákon seeks the deep privacy of the island forest to pray – an act that might rather guide our thoughts towards the Christian practice of personal prayer. But he prays to a heathen goddess, who, at the same time, would not listen to him. The reason for that might lie in the earl's earlier episodic affair with Christianity. Earlier in the saga we hear that he was baptized together with the Danish king, Haraldr Gormsson, but soon reverted to heathen practices.[20] Finally, the ultimate sacrifice is carried out, reinforcing his ties with a heathen goddess through the bond of blood. Indeed, the general image of Earl Hákon in Old Norse tradition presents him as an eager supporter of old customs – and this perception is not spoiled by his temporary association with the Christian religion.[21] The brief encounter might have rather provided him with an extra eagerness to pursue what he nevertheless considered right; as stated by *Óláfs saga Tryggvasonar en mesta* (The Greatest Saga of Óláfr Tryggvason, from around 1300): *...var hann siþan at ǫllu verri ok heiðnari. en aðr hann var skirðr.*[22]

Prímsignd/Prímsigð in *Jómsvíkinga saga* may thus be a symbolic representation of the special process that Earl Hákon was going through according to the saga, resulting in the sacrifice of his own son. The island could form a perfect setting for depicting his personal transformation from a once-baptized man to a true follower of the heathen gods again.

---

19 Blake, *Jómsvíkinga saga*, 36: "In his prayers he called upon his protector Þorgerðr Hǫlgabrúðr."

20 See, e.g., also ch. 27 of *Óláfs saga Tryggvasonar* in *Heimskringla*, volume 1; Bjarni Aðalbjarnarson, *Snorri Sturluson. Heimskringla*, Íslenzk fornrit 26 (Reykjavík: Hið Íslenzka Fornritafélag, 1979).

21 We can thus examine the imagery concerning Earl Hákon in skaldic poetry; cf. Folke Ström, "Poetry as an Instrument of Propaganda. Jarl Hákon and his Poets," in *Speculum Norroenum: Norse studies in memory of Gabriel Turville-Petre*, ed. Ursula Dronke et al. (Odense: Odense University Press, 1981), 440-458.

22 Ólafur Halldórsson, *Óláfs saga Tryggvasonar en mesta*, volume 1 (Copenhagen: Munksgaard, 1958), 147: "He then became worse in every way and more heathen than before he was baptized" (my translation).

The island setting – motifs related to magic and religion

I shall now turn from the illustrative episode in *Jómsvíkinga saga* to some other examples that highlight the significance of the insular setting in terms of connecting islands with powerful acts and transformations. A similar motif of island sacrifice and reverting to being a heathen is included in *Óláfs saga Tryggvasonar* in the 13th-century kings' saga compilation *Heimskringla*, although not in the context of the earl's battle with the Jomvikings but in relation to his earlier ventures. After a meeting during which the Danish king, Haraldr Gormsson, Earl Hákon and all his men were baptized by Bishop Poppo, the earl is said to have travelled on to some small islands off the coast of Götaland in Sweden. He then landed his ship and held a heathen sacrifice for Óðinn, possibly with the purpose of cleansing himself from the Christian religion as well as preparing for a fight ahead:

> *En er hann kom austr fyrir Gautasker, þá lagði hann at landi. Gerði hann þá blót mikit. Þá kómu þar fljúgandi hrafnar tveir ok gullu hátt. Þá þykkisk jarl vita, at Óðinn hefir þegit blótit ok þá mun jarl hafa dagráð til at berjask.*[23]

At the same time, it is of interest to point out that the place where the earl had previously been baptized, together with King Haraldr, was also an island – and indeed a clearly identifiable one. It was Mársey (that is, present-day Mors), situated in the important maritime district of Limfjord in Denmark. The same site in is sagas depicted as a meeting place for kings, a battle site and a suitable location for stationing one's fleet. The saga informs us:

> *Fóru þá menn milli þeira konungs, ok var komit griðum á ok stefnulagi. Fundusk þeir Ótta keisari ok Danakonungr í Mársey. En þá boðaði Poppó, byskup heilagr, trú fyrir*

[23] Bjarni Aðalbjarnarson, *Snorri Sturluson. Heimskringla*, vol. 1, 260. "And when he arrived at the Gauta Skerries in the east, he anchored and made a great sacrifice. Then two ravens came flying, croaking loudly. Then the earl believed that Óthin had accepted the sacrifice and that it was a propitious time to fight," [Lee M. Hollander, *Snorri Sturluson. Heimskringla. History of the Kings of Norway* (Austin: University of Texas Press, 2002), 167]. Note that *Gautasker* (in the translation "Gauta Skerries") has been interpreted as the skerries close to the Göta River off the northwestern coast of Sweden. This particular saga context in the meantime underlines the eastern position of *Gautasker* which speaks in favour of identifying them as skerries off the coast of eastern Götaland. The message similar in *Óláfs saga Tryggvasonar en mesta.* A somewhat earlier saga compilation, *Fagrskinna*, comments likewise that the earl went east to Gautland, where he threw some sacrifical sticks; cf. Bjarni Einarsson, *Fagrskinna*, Íslensk fornrit 29 (Reykjavík: Hið Íslenzka Fornritafélag, 1985), 118. A somewhat different version is, however, found in *Óláfs saga Tryggvasonar* by Oddr Snorrason (from the end of the 12th century) – in terms of extra focus on the earl's religious and political struggles; cf. Ólafur Halldórsson, *Óláfs saga Tryggvasonar eptir Odd munk Snorrason*, Íslenzk fornrit 25 (Reykjavík,: Hið Íslenzka Fornritafélag, 2006), 173-4.

Haraldi konungi, ok hann bar glóanda járn í hendi sér ok sýndi Haraldi konungi hǫnd sína óbrunna. Síðan lét Haraldr konungr skírask með allan Danaher.[24]

With regard to the scenes in *Heimskringla*, one again witnesses the combination of real features of landscape and activities that carry a deeper significance for the characters and the storyline. The parallel references in this saga demonstrate how one strategic island provides a suitable setting for conversion, whereas another one – also located along a common travel route – makes it possible for the recently converted one to find a way back to his real beliefs, providing him with the opportunity of organizing a heathen sacrifice. This particular sequence of events fits well with the understanding that sees islands as figures that reflect both isolation and interaction. The importance of interaction is demonstrated in terms of the central position of these islands – it is not simply by accident that the saga characters would come to these places. At the same time, in terms of forming an insular setting they are also somewhat separated from other territories as well as from possible unwanted company. The latter aspect may be of particular importance in the case of Earl Hákon's heathen ritual, which, according to *Heimskringla*, took place on a small island off the Swedish coast. Such a setting provided the earl with the necessary privacy, but at the same time the skerries were certainly close enough to the areas that were going to be the target for his subsequent raids.

Heathen and magic rituals that may get localized on islands can be compared to the motif that depicts islands as sites where one may encounter outlaws, criminals, and supernatural creatures. In this, the particular mixture of realistic and fantastic island features is clearly visible. The fact that extraordinary events or peculiar characters get connected with islands may be explained by the relative separation of the insular setting as compared to other territories. As the narratives show, such separation can work in several ways, either prohibiting contact between a criminal and the people he could potentially harm or, on the other hand, providing one with a hiding place from others. The dangers that may also be attached to the insular setting are explicitly expressed in terms of encounters with giants, trolls, and other creatures.[25] But as well, the simple case of a smith imprisoned on an island – and supposedly in total isolation – may result

---

24 Bjarni Aðalbjarnarson, *Snorri Sturluson. Heimskringla*, vol. 1, 259. "Thereupon messages were interchanged between emperor and king [i.e., Otto and Haraldr, my addition] and an armistice was set and meeting between them arranged to take place on the Island of Mársey. Then Poppo, a holy bishop, preached the faith to King Harald. He bore a glowing iron in his hand, and showed his hand unburned to King Harald. Thereupon Harald let himself be baptized, together with the whole Danish army," (Hollander, *Snorri Sturluson. Heimskringla*, 166).

25 A series of such events can be followed in *Grettis saga Ásmundarsonar*, see Guðni Jónsson, *Grettis saga Ásmundarsonar*, Íslenzk fornrit 7 (Reykjavík: Hið Íslenzka fornritafélag, 1936); as well as in *Orms þáttr Stórólfssonar*, see Þórhallur Vilmundarson, Bjarni Vilhjálmsson, *Orms þáttr Stórólfssonar*, Íslenzk fornrit 13 (Reykjavík, Hið Íslenzka Fornritafélag,1991). For a discussion, see K. Zilmer, "Scenes of Island Encounters in Icelandic Sagas," 238-239.

in terrifying events, which we hear of in the eddic poem *Vǫlundarkviða* (The Lay of Vǫlundr). The smith Vǫlundr, who has certain magical skills, is kept on the island of Sævarstǫð (a fictional island called "Sea-stead") by a king who forces the smith to work for him. Although Vǫlundr is supposed to be separated from everyone, the kings' sons as well as his daughter find their way to the secret island. The smith then kills the sons, and rapes and impregnates the daughter; having taken his horrific revenge, he escapes from the island, leaving behind a devastated king.[26]

As an illustration of some other fantastic island practices, I shall mention another well-known example from the eddic poetry. In the poem *Lokasenna* (Loki's Quarell, stanza 24), the trouble-maker Loki claims that Óðinn carried out shameful magic rituals (*seiðr*) on the island of Samsø (in Denmark): *Ęn þik síða / kóðu Sámsęyju í, / ok drapt á vett sem vǫlur, / vitka líki / fórt verþjóð yfir, / ok hugðak þat args aðal.*[27]

In this case, it is of particular interest to note how the name of a real and strategically located Danish island is connected to fantastic events in the framework of mythological poetry. A possible parallel to that would be a reference to another Danish island, namely, Læsø, as recorded in the poem *Hárbarðsljóð* (Hárbarðr's Song, stanza 37), which mentions a fight between Þórr and some berserk women.[28] Otherwise, island-related scenes in eddic poetry concern unidentified and/or fictional islands.

Looking at a different example derived from the saga literature, in *Eiríks saga rauða* (The Saga of Eiríkr the Red), we find a scene that speaks of certain rituals performed in an island setting by a man. When relating Karlsfeni's western expedition to *Vínland*, the saga includes a scene concerning an island somewhere along their route. The travellers come upon the island when sailing into a fjord, and they decide to call it *Straumsey* ("Stream island") due to strong currents around it. During a particularly harsh winter, the company makes its way out to the island in the hope of being able to hunt there or perhaps find a whale that has drifted ashore. A man called Þórhallr veiðimaðr (i.e., "Hunter" or "Sportsman") disappears for a while; when the others find him it turns out that he has engaged in some kind of a chanting ritual. At the same time, the rest of the company has been praying to the Christian God. As a result of the chanting, a strange-looking whale is cast onto the shore of the island; the men eat its meat, but get very sick. Realizing that the whale has been brought to them by Þórhallr's poetry to Þórr, they quickly get rid of it and once again turn to the Christian God. In this manner,

---

26 Finnur Jónsson, *De gamle eddadigte* (Copenhagen: Gad, 1932), 125-6.

27 *Ibidem*, 105: "But you once practised seid on Samsey / and you beat on the drum as witches do, / in the likeness of a wizard you journeyed among / mankind, / and that I thought the hallmark of a pervert," Carolyne Larrington, *The Poetic Edda* (Oxford: Oxford University Press, 1996), 89.

28 Jónsson, *De gamle eddadigte*, 87.

underlining the supremacy of the Christian religion, the saga now informs us that the company was able to catch fish again and survived the winter.[29]

With regard to this scene, the island in question is described as a kind of outpost at the entry to the fjord. The initial depiction also makes it clear that there were many birds on the island and that it was covered with eggs – thus showing it in a promising light, which must also be the reason as to why the travellers later decide to head out there. On the other hand, the strong currents around the island make a natural hindrance or at least add special qualities to the setting. It is perhaps not accidental that such a setting is chosen for describing the test of endurance and the competition between the powers of the old and new religion that is demonstrated to have unfolded there.[30]

Certain parallels to how particular features of landscapes can be made to fit in with the requirements of the story (or vice versa) can be drawn to an island scene occurring in the sagas concerning King Óláfr Tryggvason. The island, called Goðey, in northern Norway (Godøya, now known by the name Knapplundsøya), emerges as a heathen stronghold, since the man living there does not want to accept Christianity. Thus, in *Óláfs saga Tryggvasonar* by Oddr Snorrason (as well as in *Heimskringla* and *Óláfs saga Tryggvasonar en mesta*) a point is made about how Óláfr attempted to make his way to the island and how fierce weather conditions – caused by heathen sorcery – would not let him approach the island.[31] Naturally, with the good help of the Christian God he manages to overcome the obstacle in the end, and as a result the heathen man on the island is punished.

It is of interest that *Heimskringla* provides a separate comment concerning the district around Goðeyjar (speaking, in fact, of several islands under that same name). It is underlined that there was a strong current in between the islands and mainland. The imagery of the powerful current – which in this particular connection can actually be identified with Saltstraumen in between Saltfjorden and Skjerstadfjorden – adds its own logic to the motif of powers that would not let the king approach the island. At the same time, it is understandable why this particular island setting could have been picked out by the story tellers for the depiction of dramatic confrontations between a Christian king and a heathen chieftain. Due to the natural hindrances, this arena naturally appeared more isolated, and in such a setting it would even be natural to expect more resistance to developments that had occured elsewhere.

---

29 See Einar Ól. Sveinsson, Matthias Þórðarson, *Eiríks saga rauða*, Íslenzk fornrit 4 (Reykjavík: Hið Íslenzka Fornritafélag, 1935), 223-5.

30 The religious significance of the scene is also commented on by Geraldine Barnes, *Viking America. The First Millennium* (Cambridge: Brewer, 2001), 9-10.

31 The scene has been discussed to some more extent in Zilmer, "Scenes of Island Encounters in Icelandic Sagas", 240-3. Compare the saga motifs in Halldórsson, *Óláfs saga Tryggvasonar eptir Odd munk Snorrason*, 235-7, and Aðalbjarnarson, *Snorri Sturluson. Heimskringla*, vol. 1, 259 and 324-8.

In contrast to the latter aspect, we can point out features of island representation that show islands to be strategic outposts that provide both practical and symbolic access to wider territories. These traits connect in particular with islands located along the coasts of various mainland territories. In the kings' sagas a specific motif is to fit islands into the scheme of events dealing with the conversion of different people and territories. Earlier I referred to the island conversion of the Danish king, Haraldr. Even more significant are the island-related activities of the two Norwegian kings, Óláfr Tryggvason and Óláfr Haraldsson. Several islands along the western coast of Norway functioned as cornerstones for the kings' missionary program, showing at the same time how one could approach a wider territory through first acquiring control over its outposts. With the Norwegian coast being characterized by an outlying island belt, it is indeed natural to include the island setting into a depiction of the spread of the Christian religion. At the same time, certain islands emerge as important religious and cultic strongholds, thus contributing to the promotion of the new religion in a crucial manner.

The islands connected to the itinerary of King Óláfr Tryggvason form an illustrative case.[32] The islands that are mentioned illustrate the gradual path of the king as well as his achievements in the Christianization of the country. The ground is laid in terms of his own religious experiences in the Scilly Isles (in Britain) – according to *Óláfs saga Tryggvasonar* by Oddr Snorrason and *Heimskringla*, the king and his men were baptized there by a prophet. The later and somewhat extended *Óláfs saga Tryggvasonar en mesta* relates that Óláfr had already been primsigned/baptized earlier (while in the east), but the conversion process he had to go through in the Scilly Isles is nevertheless shown to be important. According to this saga, two men were involved in the procedure. First we hear about the hermit and his prophecy to Óláfr; in the following step, it is told that an abbot in the isles who also had prophetic skills carried out the baptism. It is underlined that Óláfr now had to serve a higher purpose; quoting the words used by the abbot in the saga:

> *Fyrir skömmu var mer synt ok sagt huerr þu uart. ok hverr þu munt uerða. þviat drottinn Iesus Kristr sendi þik af þvi híngat til eyiana. at oss hæfir at boða þer heilsamliga kenning. ok ueita þer ok þínu lidi heilagt skirnar embætti.*[33]

---

[32] It is of particular interest to mention the sequence of events as depicted in *Óláfs saga Tryggvasonar* by Oddr Snorrason; see Zilmer, "Scenes of Island Encounters in Icelandic Sagas," 239-41. In Zilmer, "On Some Pragmatic and Symbolic Features of Island Representation," additional parallels are also drawn to other sagas.

[33] Halldórsson, *Óláfs saga Tryggvasonar en mesta*, 164. "Recently it was shown and told to me who you are and who you will become. For the Lord Jesus Christ sent you in this manner to these islands here so that it would suit me to preach you the holy teaching and offer you and your men the holy baptism" (my translation).

The saga also specifies that Óláfr was 25 years old when he was baptized and that it happened in the year 993 after the birth of Christ.

In a further step, as we hear from this saga and others as well, Óláfr travelled around in the west and then made it to Norway with a clear task ahead of him. Among the Norwegian islands connected to the subsequent conversion activities are the important cases of Moster (Sunnhordland) and Selja (Nordfjord).[34] Especially the latter island holds a central significance in the tradition, as the site of the cult of St. Sunnifa. The saga material on Selja emphasizes the holy character of the island and describes the legends and miracles connected to it. Obviously, the site could then be shown as favoured by the king in the manner that suited his religious purposes, as for example, stated in *Óláfs saga Tryggvasonar* by Oddr Snorrason: *Ok at bœn byskups ok konungs ráði var þar kirkja ger ok helguð þessum Guðs mǫnnum er þar váru.*[35]

In the meantime, it can further be argued that the religious significance attached to Selja finds its own logic in the strategic location of the island; it has been underlined that the island functioned as a natural harbour for travellers in that region.[36] Parallels have also been found between this Norwegian island and some holy sites in Britain – in terms of being suitable sites where one could consciously promote a religious cult.[37] Thus, it is indeed no wonder that in the narrative representation of such island scenes, real-life strategies and symbolic motifs would often merge together.

Conclusions

This paper has focused upon certain remarkable features of the insular setting in terms of the transformative powers that are shown to be at work there. The discussed examples that connect with magical as well as religious experiences demonstrate a symbolic aspect of the narrative imagery of landscapes in Old Norse literature, which places islands somewhere in between the real and the unreal. In the scenes discussed above, islands emerge as borderline territories as well as important outposts, and they can be shown to combine in themselves the features of isolation and interaction. At the

34 In *Heimskringla* as well as in *Fagrskinna* the island of Selja is instead connected to the arrival of King Óláfr Haraldsson in Norway and called *Sæla* ("Luck"). Considering the broader cultural-historical significance of the site, see, e.g., Magnus Rindal (ed.), *Selja – heilag stad i 1000 år* (Oslo: Universitetsforlaget, 1997).

35 Halldórsson, *Óláfs saga Tryggvasonar eptir Odd munk Snorrason*, 218. "At the request on the bishop and the command of the king a church was built there and dedicated to the men of God who were there." [Theodor M. Andersson, *The Saga of Olaf Tryggvason*, Islandica 52 (Ithaca and London: Cornell University Press, 2003), 77].

36 Alf Tore Hommedal, "Bakgrunnen for helgenanlegget på Selja og staden si rolle i den tidlege kristninga av Vest-Noreg," in *Selja – heilag stad i 1000 år*, 43-76.

37 Barbara E. Crawford, "Hellige steder på De britiske øyene. Noen paralleller til Selja," in *Selja – heilag stad i 1000 år*, 160-182.

same time, it can be argued that even the obvious symbolism sometimes finds a logical reasoning in practical cultural experiences; or in the least, that the possible symbolic meanings are in a conspicuous manner made to harmonize with the general storyline. Thus, the created images are to a certain degree based upon the actual features of islands – particularly visible in the case of identifiable and strategically located islands. Furthermore, even the islands that emerge as fictional entities can have a certain narrative and cultural logic attached to them, being in many ways also motivated by the needs of the story.

In the context of religious and magical events and happenings that are connected with islands, both the dimension of separation as well as that of interaction can be illuminated. On the one hand, the insular setting can create the image of isolation and uniqueness – an aspect that is emphasized in terms of special rituals, deep spiritual experiences, and extraordinary encounters. On the other hand, islands can emerge as cornerstones and important outposts, providing access to other territories – in the case of islands located along strategic travel routes, the communicative openness is obvious, which also turns them into suitable sites for introducing new impulses which can then be passed on to other regions. In this, the fusion of island symbolism and pragmatism can again be noted. It would therefore be important to keep in mind both of these perspectives in further studies, for example, when analyzing the various features of island representation in relation to the overall manner in which real and cognitive landscapes are formed in Old Norse narrative sources.

MONASTIC "ISLANDS" IN MEDIEVAL DENMARK: INSULAR ISOLATION IN IDEAL AND PRACTICE

Johnny Grandjean Gøgsig Jakobsen

Every now and then, when I am asked to describe the medieval monastery, there are several ways to do so. One of them is to make use of the topic of these conference proceedings: *Islands!* Islands isolated in more than one sense. For medieval *religiosi* – the monks, canons regular, friars and nuns – the monastery (or the abbey, priory or nunnery) constituted a sort of haven, an enclosed area within this world which rightfully belonged to another world. For my students or when addressing an audience at an open lecture, I usually make a comparison with present-day embassies, physically located in one country but symbolically and jurisdictionally belonging to another. In the same way, medieval monasteries can be seen as embassies of Paradise within the temporal world. When you passed the gate and entered the enclosed area behind the walls, you were – at least ideologically – no longer in the local county, but in some sort of forecourt of Paradise. Here the rules of your local village, count or king were second to the rules of God, represented by the constitutions of the religious order in question..

But perhaps more importantly, the monastery was a refuge from the Devil and all his doing in the temporal world, at least ideologically, since powerful mobs and magnates at all times have dared to ignore the godly primacy and jurisdictional sanctuary of monasteries. At other times, at least as often portrayed in anti-monastic traditions, the inhabitants of the monastery themselves, misled by evil thoughts and longings, invited all the sins and dangers of the outside world into their sacred house. Often this is also described in monastic sources of the Middle Ages, where a reformist fraction of the order takes it upon itself to clean up the ungodly mess allowed by former generations. For instance, when Bishop Absalon put the French Augustinian, William of Æbelholt, in charge of the small convent of canons regular on the islet of Eskilsø outside Roskilde in 1165, he found a most unholy monastery, lost to the Devil for years. The canons: "...dozed in disgraceful idleness, they ate and drank to abundance in a royal way in arrogance and wrong living." Just in this sentence alone, the chronicler can list no less than three mortal sins, which were, of course, imposed on the canons by the Devil himself. When William began to reform the convent and bring them back to the right course, the Devil became so angry that he made the canons try to murder the reformist abbot on several occasions, for instance, by lighting a fire next to William's bed at night. All of this was in vain, however, and William forgave his brethren because he knew they were not acting for themselves, but on behalf of the Devil.[1]

On other occasions, it took a whole new order to reform the fallen monasteries of an older order. This was allegedly the case for several of the Cistercian abbeys in Denmark, which, according to Cistercian chronicles, were often based on fallen Benedictine convents in severe need of reform.[2] Just like a military fortress of the temporal world, the religious fortress of the monastery was no stronger than the garrison that protected it. If the guard was weak and lazy the forces of Evil could easily climb the wall, sneak through the gates, and so take command of the entire monastery. But if the garrison was well trained, disciplined and alert, the Devil and his army could not enter this forecourt of Paradise. Indeed, such use of militaristic language and metaphors are not just my invention when describing medieval monastic life; contemporary chronicles and letters are full of them. For instance, numerous militaristic metaphors can be found in the preserved letters of Abbot William, mentioned above.[3] The idea of the cloister wall and gate being the checkpoint between our two worlds is shown among other places in a letter from the late 12th century, where William describes how he can actually see the devils sitting on the wall above the monastery gate, just waiting for a chance to sneak in. In less dramatic phrases, a large part of *The Rule of St. Benedict* deals with the same problem, of how to prevent the dangers and the disturbances of the temporal world from entering the monastery and disturbing the sacred peace, and why the monks should stick to their *stabilitas loci* and avoid leaving the safe place of the monastery – to stay on their guard, so to speak. Or on their island.

So, in short: In the fluctuating and wild ocean of the temporal world, full of changes and dangers, where your ship could so easily be wrecked and you yourself could so easily drown or get devoured by dreadful monsters, in this unkind ocean, the medieval monasteries constituted a group of heavenly islands, where peace and God's will ruled for all eternity, where the walls were like the shores protecting you from the Sea of Sin, and where you could concentrate on a divine and apostolic life, in solitude and

---

1 *Wilhelmi Abbatis Vita*, published in a Danish translation by Hans Olrik in *Danske Helgeners Levned* (Copenhagen 1894; reprint Copenhagen: Rosenkilde og Bagger, 1968), vol. 2. Quotation translated from p. 190.

2 For Sorø Abbey, *Liber Donationum Monasterii Sorensis*, published in *Scriptorum Rerum Danicarum* vol. 4, with a Danish translation by Svend Ranvig in *Sorøkrøniken*, vol. 2 (Copenhagen: Finn Jacobsen, 1986). For Øm Abbey, *Exordium Monasterii Carae Insula*, published in a Danish translation by Jørgen Olrik in *Øm Klosters Krønike* (revised edition: [Skanderborg]: Scriptorium, Øm Kloster Museum 1997).

3 *Epistolæ Abbatis Willelmi de Paraclito*, published in *Danicum Diplomatarium* 1. series, vol. III (with Danish translations in *Danmarks Riges Breve*, same series and volume).

enjoying the protection and help of your fellow brethren or sisters of the convent and the qualified guidance of the abbot, prior or mother superior.

Thus, medieval monasteries can be seen as isolated islands in a mental sense, but to some extent several orders and convents tried to take this ideological concept of the monastic island further, in an actual and physical-geographical sense as well. Especially in the Early Middle Ages and the beginning of the High Middle Ages, several monasteries of Benedictine and Augustinian origin were founded in Northern Europe on small, remote, and isolated islands. For historians working with the Viking period, the Benedictine Priory of Lindisfarne on the coast of Northumbria is a famous example; during the Second Muhu Conference we needed look no further than our own beautiful, historical surroundings at the Augustinian Utstein Kloster on Klosterøy. Certainly, the idea of placing a monastery on an isolated island fitted well with the ideology of early monasticism, but to what extent was it actually carried out in medieval Denmark?

As one can see on the monastic map of medieval Denmark in Figure 1, the country is in large part constituted of islands, but islands like Sjælland, Fyn, Lolland, Falster, Rügen, and Mors can hardly be termed "isolated," in fact, they were some of the most densely populated areas of medieval Denmark. Perhaps more interestingly, the smaller islands of the Danish kingdom are actually characterized by an almost complete absence of medieval monasteries. In Denmark, the religious orders of the Middle Ages found rich opportunities to settle on small and fairly remote islands, but they very rarely seem to have done so.

On the map (fig. 1), I have depicted the 137 known monasteries of medieval Denmark. The different colours and shapes of symbols refer to the different orders, as explained in the map legend. Going through the entire list of monasteries, I have identified only six which can be said to lie on some sort of isolated island with no other settlement. Three of these monastic islands are coastal, whereas the other three are "inland islands" situated on small islets in fresh-water lakes. In the case of Bosjö, it can be argued that the Benedictine nunnery is indeed only placed on a peninsula in Lake Bosjö, not on a real islet, and in the cases of Gavnø and Eskilsø it is difficult to claim that they were all that remote and isolated, as they are situated extremely close to the important and easily accessible cities of Næstved and Roskilde. Even so, the Augustinian convent of Eskilsø, which was the one Abbot William was sent to reform in 1165, was deserted after only ten years, as William decided that it was impossible to run a proper abbey on a small, isolated islet in the Fjord of Roskilde. With the blessing of the bishop the whole convent was moved to a new site, Æbelholt, in inland Sjælland. William's completely unromantic and un-idealistic, but very pragmatic, decision was probably quite characteristic for Danish monastic foundations in general: Small, isolated and remote islets were generally not well-suited for establishing a monastery in the "modern monasticism" of high and late medieval Denmark. Either the monastery was dependent on some sort of contact with its estates, which was all the more difficult from a small island, or – in the case of the estateless mendicant priories – the brethren of the convents needed to be in close contact with the surrounding society in order to perform their stated goal (i.e., preaching) and, indeed, to survive.

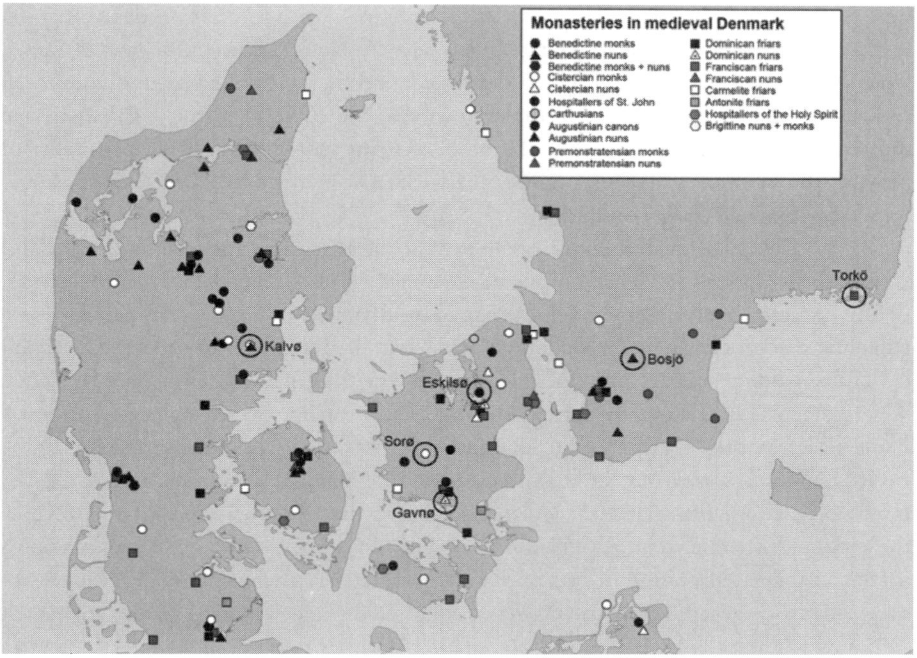

Fig. 1: Distribution of monasteries in medieval Denmark, pointing out six monasteries located on small islands with no other settlement.

In none of the cases of monastic islands in Denmark did the religious orders choose the locations themselves. In Kalvø, we do not know the background for the original Benedictine house, which was probably only used as a remote cell of contemplation for the monks from the nearby abbey of Veng,[4] but when the Cistercians took over Veng Abbey and were moved from there to Kalvø, they only stayed for about four years before they abandoned the place for a new foundation in Øm.[5] Their fellow brethren in Sorø probably would have preferred to do the same, but they were prevented from doing so by the fact that the existing Benedictine church on the island of Sorø already held the graves of the forefathers of the donating family, the Whites. For the Cistercians especially a monastic location on a small freshwater-lake-island was far from

---

4 Tore Nyberg, *Monasticism in North Western Europe, 800-1200* (Aldershot: Ashgate, 2000), 109-12.

5 *Exordium Monasterii Carae Insulat*, ch. 19-21.

ideal, since their monasteries more than all other orders of their days were focused on the necessity of clean, running water, which is hard to obtain on such locations in Denmark.

For the later orders also, small island locations were not a choice of the orders themselves. When the Dominican nuns received Gavnø as their residence in 1403 it was by a decision by Queen Margaret, who herself wanted the site Gladsaxe in Skåne, which had been chosen originally for the foundation by the donor and the Dominican order.[6] And in the case of Torkö, the closest thing we come to a true isolated, coastal monastic island in medieval Denmark, the Franciscan friars really had no possibility of moving the convent elsewhere, as it was founded to service an already-existing and popular pilgrimage chapel on the site.[7]

Thus, so far, the conclusion seems to be that monastic islands in medieval Denmark were more an ideal than practice in an actual physical geographical sense. Still, before completely dismissing the implementation of an island ideology, it may be worth enlarging the concept a bit by looking at actual islands not just in an oreo-hydrographical understanding, but also as possible "islands" in the landscape in a broader sense. Whereas the Cistercians experienced the concrete problem of getting water on small Danish isles, it was apparently quite an important part of the Cistercian self-image of the twelfth century to seek out the most remote and uninhabitable places for their monasteries. Cistercian chronicles and regulations, such as the *Exordium parvum*, speak of how the monks deliberately chose "...places of horror and vast solitude...," which may be a slightly exaggerated version of one of the many Cistercian myths, but even recent historians tend to accept that the White Monks did indeed found their abbeys in "...uninviting areas outside manorial boundaries, lands that no one else cared to till."[8] In Denmark, this claim of the Cistercians to cultivate the desert is – just like in most other North European places – rather difficult to maintain when taking a closer look, since the Cistercians in most places seem to have preferred rather well-cultivated and populated areas, indeed, in several places they seem to have created their "remote deserts" themselves by closing and tearing down already existing villages on their chosen sites. An example of this can be found at the Abbey of Esrum in the northernmost part of Sjælland, where a number of villages can be accounted for in the vicinity of the abbey, all of which were apparently closed down and deserted shortly after the monastic foundation.[9] Quite close to Esrum is another example of what appears to be "too much

6 *Diplomatarium Danicum* 4. ser. VII no. 259.

7 Jørgen Nybo Rasmussen, *Die Franziskaner in den nordischen Ländern im Mittelalter* (Kevelaer: Verlag Butzon & Bercker, 2002), 109-11.

8 Louis Julius Lekai, *The Cistercians - Ideals and Reality* (Kent, OH: The Kent State University Press, 1977), 282.

9 Brian Patrick McGuire, "Property and Politics at Esrum Abbey: 1151-1251," *Medieval Scandinavia* 6 (1973): 132-4; idem, "Esrum Kloster - forhistorien og grundlæggelsen," in *Bogen om Esrum Kloster*, ed. Søren Frandsen, Jens Anker Jørgensen and Chr. Gorm Tortzen (Frederiksborg Amt, 1997), 22.

remoteness" for a monastic settlement. In the rather un-inviting district around Asserbo, north of Arresø, Archbishop Eskil invited the Carthusian Order to establish a monastery in 1163, but the monks of this otherwise quite ascetic society, who were sent from France to Denmark to be the pioneers of the foundation, decided to leave again after only seven years, telling their superiors that this was simply not a place suited for human settlement – not even for Carthusians![10]

With that said it is not completely fair to dismiss the monastic idealisation of islands and deserts as ideals only with no foundation in Danish medieval reality. Surely there is a striking absence of medieval monasteries on the smaller Danish isles and in the western part of Central Jylland, which are also the most thinly populated areas of the kingdom. Still, it cannot be denied that a surprisingly high number of Benedictine nunneries in Northern Jylland are not in completely deserted and uninhabited places, but still rather remote locations where it can be hard to explain the choice of site with anything else than a taste for solitude. The same can be said for the Premonstratensians in central and eastern Skåne.

The location of monastic foundations is, of course, initially determined by the outside founders of the convent – be they royal, lay or secular-ecclesiastical – but as can be seen in some of the examples here, supplemented by several others from the Danish as well as foreign material, monastic orders generally felt little obligation to stay put in a place which they did not like![11] The Franciscans of Malmö changed the location of their priory several times before finally finding the right place for it within the city area,[12] whereas the Dominicans of Helsingborg refused to move their priory even though this was demanded by the king – apparently, they were quite happy with their location.[13] Seemingly the same can be said for the Dominicans in Bergen, who withstood centuries of harassment from the neighbouring canons at the cathedral chapter, who terrorized them in all sorts of unholy ways to make them move away, but the Black Friars stubbornly stayed put.[14] In Sweden, the first Dominicans trying to settle in Sigtuna had to flee due to opposition from the archbishop. They were instead given a site in nearby Sko by a potentially important political ally, Knut Holmgersson Långe, who later became the Swedish king, but as the place did not suit them, they simply decided

10 Jens Anker Jørgensen and Bente Thomsen, *Gyldendals bog om danske klostre* (Copenhagen: Gyldendal, 2004), 148.

11 Johnny Grandjean Gøgsig Jakobsen, "Prædikebrødrenes samfundsrolle i middelalderens Danmark," PhD diss. (Odense: University of Southern Denmark, 2008), 190-1.

12 Rasmussen, *Die Franziskaner in den nordischen Ländern*, 508; Jörgen Kling, "De religiösa institutionerna i Malmö," in *Liljan - om arkeologi i en del av Malmö*, ed. S. Larsson (Stockholm: Riksantikvarieämbetets förlag, 2006), 111-2.

13 *Diplomatarium Danicum*, 3. ser. VI no. 120; Jakobsen, "Prædikebrødrenes samfundsrolle," 217.

14 Ibidem, 153.

to leave Mälardalen completely for the time being.[15] A similar tale can be told from England, where the first Dominicans were offered sites in both Canterbury and London by the archbishop and the king, but the friars politely turned them down, at least for the moment, as their superiors in the order had told them to go to Oxford.[16] All this just goes to show that medieval monastic convents were far from being as humble and submissive to the wishes of local magnates as some modern historians seem to think.

Remote locations or "isolated islands in the medieval landscape" did apparently correspond better with some monastic orders than with others, in Denmark especially with Benedictine nuns and Premonstratensian canons regular, and to some arguable extent also to Cistercian monks. Exactly why this was so deserves more looking into – more than I can give here and now.

Instead, I am going to end my paper by turning the focus once again to the order of my primary attention for several years now, the Dominicans, who, just like the other large mendicant order, the Franciscans, differed from the old religious orders by not obeying the general monastic rule of *stabilitas loci*. In fact, the main purpose of the friars was to mingle as much as possible with the rest of society in order to make people understand the true meaning of being a Christian, which for the Dominicans, or Friars Preachers, as they were rightfully called in the Middle Ages, was first and foremost done by preaching.

As I have tried to show in my doctoral dissertation, however, the homes of the Dominican Friars Preachers themselves, their priories, also differed a great deal from regular monasteries as they were far from being isolated from the rest of society. Firstly, all, or almost all, mendicant priories – especially Dominican priories – were located in the most densely populated places of medieval society, the big cities. This was meant to bring the Dominican friars and preachers as close to as many people as possible, but it was also in order to bring people close to the home of the friars. Mendicant churches not only allowed access for a lay audience; the convents were highly dependent on such lay traffic in their churches, attending canonical hours, masses, sermons and other services as well as burials.[17] Not only the churches, but also the rest of the priories were often open for lay use. The main halls of the priory were used surprisingly often for various kinds of lay meetings, meetings which had nothing at all to do with the friars themselves. The reasons for this are probably manifold, but one important reason seems to be that the priories held some sort of neutral position for the parties of the meetings concerned. Once again, their island-like status in society apparently made

---

15 *Historia ordinis predicatorum in Dacia*, published among others by J.G.G. Jakobsen with Danish translation (Centre for Dominican Studies of Dacia, 2007: http://jggj.dk/HOPD.htm).

16 Clifford Hugh Lawrence, *The Friars – The Impact of the Early Mendicant Movement on Western Society* (London and New York: Longman, 1994), 131.

17 Jakobsen, "Prædikebrodrenes samfundsrolle," 116-37.

Dominican and Franciscan sites neutral ground for the parties involved.[18] This is not to say that there was free traffic continuously going in and out of the priories, far from it. Just like Benedictine, Cistercian, and Augustinian monasteries, the mendicant priories were surrounded by walls and had gates guarded by alert doorkeepers, but the mendicant priory was not such a secluded and isolated place of peace and contemplation as the monasteries of the old orders.

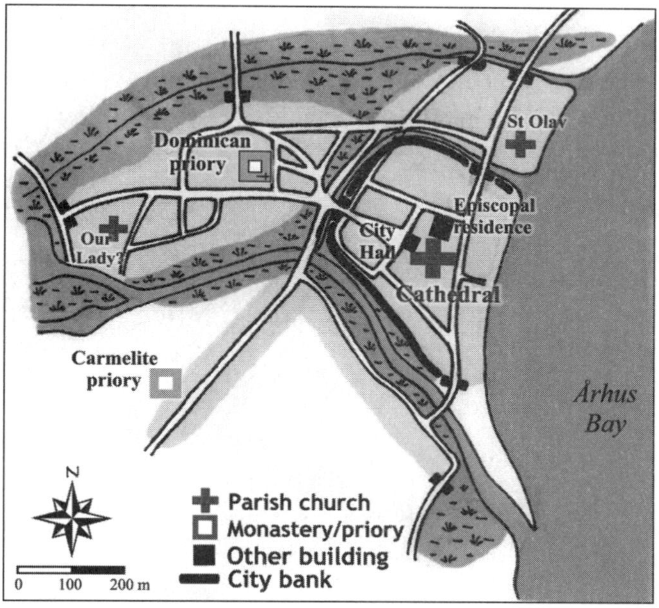

Fig. 2: Reconstructed city map of Århus showing the location of the Dominican priory in the northern part of town, a location which made it most convenient for the friars to place the priory church south of the cloister to make it as easily accessible as possible for urban residents.

This semi-openness towards the rest of society constituted a continuous challenge for mendicant priory planning. Within the priory site, the priory itself would be located as close as possible to the centre of town and the trafficked streets in order to get the priory, and especially the church, as close to the potential audience as possible. In fact, in cities where the priory was situated in the northern part of town, the mendicant friars were always exempted from the general monastic rule of putting the church north of the cloister; in half of the mendicant priories in medieval Denmark, the church, as in Århus (Figure 2), was situated south of the cloister.[19] This location not only put the priory church as close to the lay population as possible, it also offered a

18 Ibidem, 204-15.

19 Ibidem, 193-5.

kind of protection for the rest of the buildings – the chapter hall, the library, and the dormitory – from the noises and smells of the medieval city streets. Contemporary sources from Germany, Flanders, and France relate that this was actually a highly considered issue in mendicant priory planning.[20] Furthermore, the adjacent gardens of the priory in its immediate hinterland constituted a "peaceful, heavenly island" protected by the church and the domestic buildings from the disturbances of the urban centre. In fact, even today when I enter a Dominican cloister or such a protected priory back garden – be it in Ribe, Århus or Rome – I am always struck by its surprising peacefulness even at the otherwise noisiest times of day in its immediate urban surroundings. In this sense, even a Dominican priory in the middle of a large medieval city did indeed constitute a peaceful, isolated island in the middle of an agitated, temporal ocean.

20 Ibidem, 189-91.

RELIGIOUS ATHLETES – ON THE PERCEPTION OF THE BODY IN MEDIEVAL ASCETICISM

Torben Kjersgaard Nielsen

And I feel your fist
And I know it's out of love
And I feel the whip
And I know it's out of love
And I feel your burning eyes burning holes
Straight through my heart
It's out of love

I accept and I collect upon my body
The memories of your devotion
I accept and I collect upon my body
The memories of your devotion.[1]

This article will not deal with sports or athletics in the modern meanings of these terms. Rather, I propose to consider religious men and women of the Middle Ages as using the capacities and strengths of their human bodies to achieve a religious goal, almost like professional athletes and sports people do nowadays. I aim to show how the body was in fact a decisive "target area" for religious achievements. In their own understanding, the religious athletes would – through different kinds of asceticism and numerable kinds of bodily mortifications – work to bring themselves nearer to God. The playground for this, the athletic arena if you will, was the human body. In this process, the human body was seemingly violently denigrated and made low in a rather conscious process of disparagement. From another viewpoint, however, the human body was also revered and honoured, in a way the *sine qua non* of these spiritual athletics. I shall attempt to show examples of this in this article.

---

1 Excerpt from "Fistful of Love" by Antony & The Johnsons (lyrics: Antony; music: Antony and the Johnsons) (from the CD: *I am a Bird Now*, Roughtrade Records 2004).

The famous medieval classificatory system of *laboratores*, *bellatores*, and *oratores* was installed by the *oratores* themselves.[2] The system was, of course, refined endlessly over the course of the Middle Ages, but one aspect remained in the numerous classifications available: Ecclesiastics and the men and women religious always stayed at the top of these hierarchic classifications – closest to God. Any social or religious classificatory system of a hierarchic nature, however, involves some sort of marginalisation or exclusion. What was special about the religious classificatory systems of the High Middle Ages was that the minorities excluded in these systems were the religious groups and persons themselves. They were marginalised and isolated at the pinnacle of these systems, not at the base as minorities normally are.

It is thus fair to say that the religious minorities, themselves the authors behind the classificatory systems, were successful in at least two ways: They excluded themselves in exalted and isolated positions in these classificatory systems, and at the same time established their own choice of life as the ideal for the whole of society. In this sense, the religious men and women in the High Middle Ages were isolated and excluded, but at the same time also beloved and honoured, since their isolation and exclusion was of their own choice and it was believed to be a common good for all members of society.

In the central Middle Ages, furthermore, a number of new religious movements were born, and with them also new spiritual achievements and ideas. The new religious movements no longer saw themselves only as intercessors asking God for mercy on behalf of a sinful mankind and thus working socially for the salvation of man. In this period the "inner," "personal," or "spiritual" growth of the religious himself became important.[3] The monastery or the hermitage was acknowledged as the school of life, in which the religious athlete would be offered the opportunity to commit himself to moving closer to God through the use of deliberate methods developed exactly for such purposes. Behind this lay a simple – Augustinian and Neo-Platonic – idea, that the religious athlete, by using different self-sacrifices, readings, prayers, meditations, and body-controlling techniques, and progressing up and in a forward-going movement, would work himself towards exactly that likeness to God which was thought to have been lost with the Fall and Original Sin.

The ninth prior in the monastery of La Grande Chartreuse, Guigo II (d. 118 authored a small treatise – the *Scala claustralium* or *Monk's ladder*. Guigo's treatise under-

---

2 Cf. the brilliant chapter: "The Orders of Society" in Giles Constable, *Three Studies in Medieval Social and Political Thought. The Interpretation of Mary and Martha, the Ideal of the Imitation of Christ, the Orders of Society* (Cambridge: Cambridge University Press, 1998).

3 Cf. the many interesting contributions to this in Robert L. Benson and Giles Constable (ed), *Renaissance and Renewal in the Twelfth Century* (Oxford: Clarendon Press, 1982) and Giles Constable's magisterial work: *The Reformation of the Twelfth Century* (Cambridge: Cambridge University Press, 1998).

lines this new monastic idea of ascension and continuous development toward the desired likeness to God. Important elements in this development of a truly religious way of life were, of course, traditional remedies like reading, meditation, prayer, and contemplation. What strikes me as new, however, is his assertion of the notion of progression and personal endeavor now believed to be inherent in these monastic practices. Guigo relates to this by using the sensory regimes of the human body to allegorise the necessary stages in personal spiritual development. His use of a specific "bodily" language thus serves the point of appropriating the Truth of God. Just as the body digests food for it to survive – at times even enjoying the taste of well-prepared food – so also the wisdom of God must be disintegrated and "chewed" to be properly "digested" by the mind:

> Reading seeks for the sweetness of a blessed life; meditation perceives it; prayer asks for it; contemplation tastes it. Reading, as it were, puts food whole into the mouth; meditation chews it and breaks it up; prayer extracts its flavor; contemplation is the sweetness itself, which gladdens and refreshes. Reading works on the outside, meditation on the pith; prayer asks for what we long for, contemplation gives us delight in the sweetness which we have found.[4]

In the high Middle Ages, however, not all writers would have agreed to the positive use of the human body in allegories like these. The religiously awakened and spiritually active, monks, hermits, *reclusae* and whatever odd name they were given or they themselves took, were often admired particularly for their religious "athletic" exercises. It was perceived that their way of life in fact did some good for *all* the sinners of this world.

In applying exactly these activities, however, the religious athletes marginalised themselves in a way, as stated above. These exercises were not everyday practices for every man. In these spiritual practices a harsh, but still well-known and traditional, differentiation between the noble mind and the ignoble body was violently pursued by some of these religious athletes. The heavenward soul and its achievements were elevated, while at the same time the earthbound body and its functions had to be negated. It is quite easy to find written statements from the period uttering a seemingly fundamental contempt for the human body. A work by Cardinal Lothar of Segni, later Pope Innocent III, provides a lucid example. In his *De miseria humanae conditionis* ("On the Misery of the Human Condition") from the 1190s, Cardinal Lothar characterises mankind negatively as being of a purely bodily – and hence despicable – nature:

> For sure man was formed out of earth, conceived in guilt, born to punishment. What he does is depraved and illicit, is shameful and improper, vain and un-

---

[4] Guigo II, *The Ladder of Monks and Twelve Meditations*, translated with an introduction and notes by Edmund College and James Walsh, Cistercian Studies Series, vol. 48 (London: Mowbrays, 1978), 82-3.

> profitable. He will become fuel for the eternal fires, food for worms, a mass of rottenness. ...I shall try to make my explanation clearer and my treatment fuller. Man was formed of dust, slime, and ashes; what is even more vile, of the filthiest seed. He was conceived from the itch of the flesh, in the heat of passion and the stench of lust, and worse yet, with the stain of sin. He was born to toil, dread, and trouble; and more wretched still, was born only to die. He commits depraved acts by which he offends God, his neighbor, and himself; shameful acts by which he defiles his name, his person, and his conscience; and vain acts by which he ignores all things important useful, and necessary. He will become fuel for those fires which are never hot and burn forever bright; food for the worm which forever nibbles and digests; a mass of rottenness which will forever stink and reek...[5]

The human body, created as it was by depraved acts, would continue to act in depraved and shameful ways. It was against this body that the religious athlete must struggle. The religious athletes could thus be perceived as soldiers taking up a fierce battle against what was perceived a treacherous human body in which evil sin lurked.

In his vita from the early years of the thirteenth century one reads how a worldly knight, William, paid St. Wulfric a visit. Upon witnessing Wulfric's holy endeavors in his self-chosen solitude, William decided to offer Wulfric his armor. This gift from the knight would now give the author of the vita a perfect opportunity to juxtapose the choices of life in the two persons, the knight and the recluse, and to praise St. Wulfric for the higher purpose in his spiritual choice of battles and his truly religious art of warfare:

> The Lord William ... paid a reverent and hurried visit, pressing his anxious hope that he might share in the warrior's triumph by furnishing the arms. He handed over his own hauberk, as though to a mightier knight than he, and dedicated this piece of military equipment to the use of God's recruit. And so the solider newly harnessed advanced to the spiritual battle, striking terror into the foe by turning the arms of the world into weapons of righteousness. Thus he did equip himself for daylight warfare. At night he plunged himself naked in a tub of cold water, and there Christ's little lad would offer his master all the psalms of David, singing to Him with drum and psaltery. Thus after each new baptism he bloomed afresh in the splendor of innocence and, like a flock of shorn ewes that have come up from the washing, emerged from the water whiter than milk. Putting back on his shirts of hair and mail (noble both, but

[5] Quoted from Marvin Perry (ed.), *Sources of the Western Tradition*, vol. 1: *From Ancient Times to the Enlightenment* (Boston: Houghton Mifflin , 1995), 277-8.

> the second emphatically nobler), he spent what remained of the night office in hymns and prayers, punctuated with frequent genuflections.[6]

In many of the writings about the tough sport of the religious athletes, food and the consumption of food take up an important place. More or less controlled and self-induced deprivation of even simple food, otherwise thought necessary for restoring the body, was often regarded as a creditable strategy for bodily control. The vita of the prominent Cistercian, Aelred of Rievaulx, underlines both the respect and the worry that sometimes characterised the public when witnessing the often harsh ascetic practices by the religious athletes. Obviously – and this is exactly one of the points the author of the vita wished to drive home to his readers – that such a degree of bodily control could only be reached through many years of devoted practice and deep religious zeal. The story in Aelred's vita is furthermore interesting in that it underlines some rather complex relations of insecurity and (a longing for) recognition on the one side, and – on a slightly different level – of communality and isolation on the other. At the same time, the text works with strong metaphors for sickness and recovery. When medical categories are involved in texts like these, doctors and an authoritative discourse on health and wellbeing are also often invoked. By posing himself in direct opposition to his doctors, Aelred in this text displays an almost subversive attitude, thereby highlighting yet other important questions – especially in his Cistercian setting – of obedience and authority.

The quote below from Aelred's vita is lengthy, but in my mind worthwhile. It testifies that the religious athlete in fact would have been a troublesome character in many monasteries. The religious athlete would display disconcerting and perplexing features of excessive spirituality through his own body for all to see, but – and this is the point that might have made him admired yet troublesome in a monastic setting – not for all to copy:

> Throughout those four years before his death, our father experienced what I may call a second circumcision, not by the removal of superfluities which even did not exist, but by depriving himself of necessities very helpful to him in his weakness. He made his little body free of everything that is pleasant in this present life. He sacrificed himself on the altar of unfailing suffering: hardly any flesh clung to his bones; his lips alone remained, a frame to his teeth. The excessive emaciation of his body and the thinness of his face gave an angelic expression to his countenance. Eating scarcely anything and doing less, by his unbelievable fasting he lost altogether, and no wonder, the desire for food. During this period he rejected the curatives which he had been wont to take, and if by chance he tasted anything of that kind in his mouth he took it out

[6] Pauline Matarasso (ed. and tr.), *The Cistercian World – Monastic Writings of the Twelfth Century* (Harmondsworth: Penguin, 1993), 239.

> with his fingers and, while his attendants were engaged on other things, threw it on the ground and ground it to powder with his foot so that it should not be seen. Nor, as had been his custom, would he take a little wine, but would have it watered and sip, rather than drink, a mixture which was more like water than wine, although the physicians had prescribed pure wine as the particular remedy of his infirmity, asserting that otherwise the disease would take its course and that he would soon die. But the father, taking his soul in his hands, gave greater weight to his own counsels than to those of the physicians, and for God's sake despised the cure of the body and considered in alle ways the health of the soul.[7]

Paradoxes abound in such stories of protracted and, to be sure, enervating, but mindful and attentive bodily techniques. As mentioned earlier, a part of the spirituality of these zealously religious was concerned with the proper way to establish a lost likeness to God. A part of this movement towards God was believed to consist in the ability to rid oneself of one's own corporeality. In this endeavor, a clear contrast would often be established between the rejection of corporeality on the one side, and the careful nursing of the same corporeal body on the other. This is a dichotomy visible in many texts. A number of these texts served to introduce to their readers persons who would serve – more or less directly – as models for imitation. One of several vitae of the most famous Cistercian, St. Bernard of Clairvaux, stated:

> Further is he hardly ever driven to eat by the desire for pleasure but merely by the fear of becoming too weakened. Because when he must eat, he is satisfied by the thought of food alone, before he even eats. He approaches his food as if a malady. From his first year, counting from his new life, he is used to throw up undigested everything he takes in, since he has always been of a too fragile and delicate bodily constitution, and his stomach is destroyed by too much fasting and vigils, by cold and by work, by very hard and persisting labors. If any of his food will be digested in a natural way and is sent into his interiors, then these parts of him will be tormented by painful diseases, and only in great pains will he give if off again. If anything of it stays in him, it will be to nurture his body. However, this does not so much serve the purpose of keeping him alive as it serves to postpone his death.[8]

The Benedictine abbot, later bishop of Chartres, Peter of Celle (d. 1183), in a small treatise, *De afflictione et lectione* (On suffering and reading), describes the necessity and usefulness of bodily mortifications. In his text, Peter of Celle exhibits some interesting metaphors, which underline the cognisant ideas about the body embedded in many such

---

7 Walter Daniel, *The Life of Aelred of Clairvaux*, trans and annot. by F. M. Powicke, Cistercian Fathers Series 57 (Kalamazoo: Cistercian Institute, 1994), 127-8.

8 Guillelmus S. Theoderici, *Liber Primus*, in *Patrologia Latina* 185, col. 239B-240A.

texts. In this text, the human body is negated and humiliated to the extent that it becomes a commodity to be sold and bought. Another often recurring theme is at play in his text as well, namely, the idea that the body in fact was incomplete. This idea seems to imply that the body can only become complete through mortifications and discipline. It is exactly in this way that the human is important to Peter of Celle. To him, the body is a formable element which must be manipulated. Only with manipulation can the body be considered of value after all:

> A room has one of two qualities, depending on the way of life of those, who dwell in it. It is a hard place for carnal people, but a pleasant one for spiritual people. It is a prison for the flesh, a paradise for the mind. It is a market where the butcher sells small or large amounts parts of his flesh to God, who comes as a customer. The more of his flesh he sells, the greater grows the sum of money he sets aside. Let them therefore increase their wealth and fill their purses by selling their own blood and flesh, for flesh and blood will not possess the kingdom of God. Voluntary mortification of the flesh is the redemption of man's soul; which is the riches of the saints. The Lord says, 'He who looses his souls will find it' [Matt. 10:39] Although what the Apostle says is true: 'I do not find good in my flesh' [cf. Romans 7:18] nevertheless I can make a good profit from the flesh and acquire glory for it, if I remove wantonness. Yet, the flesh is childish and extravagant, unless it is restrained by the yoke of discipline and granted no independent management of itself and its belongings.[9]

A traditional negation of the body is, of course, apparent in this text. Not many qualities immediate to the body are discernable in this yearning for mortification; the body must be subdued and suppressed. To Peter of Celle, the incomplete body can be made complete, not by adding what is presumably lacking, but rather by taking away what is essentially bodily. If the body had a purpose, then, it was as a tool of its own destruction.

However, this was not the only available attitude towards the body among the religious athletes. Other, supplementary, attitudes existed. The religious athletes in medieval society formed a social group which by their choice of life was both honored and marginalised. In this respect, they can actually be likened to another marginalised figure, despised and isolated but loved at the same time, Christ himself. The central Middle Ages abound with religious persons referring to the example of Christ, either symbolically and allegorically or in direct and physical ways. Thus, the notion of *imitatio Christi* is an important feature of the spirituality of the High Middle Ages. An example of this *imitatio* strategy is available in Peter Damian, who wanted to suffer for the sake of Christ and who mortified his flesh, allegedly because he could not become a martyr since in his

---

[9] Peter of Celle, *Selected Works*, trans. Hugh Feiss, Cistercian Studies Series, vol. 100 (Kalamazoo: Cistercian Institute, 1987), 140-1, and quote from p. 132.

lifetime the age of persecution in the early Church was long gone. The prominent scholar on medieval spirituality, Giles Constable, characterises Peter Damian as being "his own executioner and a self-martyr, who willingly imposed suffering and even death upon himself."[10]

Damian not only practiced an ascetic lifestyle, he also wrote admiringly of others. Peter Damian authored a vita of an Umbrian ascetic, a certain Dominic Loricatus (d. 1060), another fashionable religious athlete. Loricatus was known to carry seven or eight iron plates (*lorica*) tied to his body around his waist, neck, and legs. Peter Damian says of Loricatus that he would pray for long periods with his arms outstretched while at the same time performing numerous penitent genuflections. The performance of such exercises seems to come close to the modern idea of athletics.

Loricatus is mainly famed for his self-flagellation. According to Damian, Loricatus would strike himself while reciting the Psalms. In this exercise he would offer himself a thousand blows for every ten psalms and carry out a hundred genuflections for every fifteen psalms. A quick calculation makes this amount 15000 blows and a thousand genuflections for a full rendering of the Psalms. It was reportedly his habit to recite the Psalms twenty times in six days, and once he even reached nine full renderings in a single day. I will leave the math to the reader, but what is certain is that Loricatus would have to have been in what is nowadays called very good shape to do this.[11] Dominic Loricatus is, furthermore, said to have carried stigmata on his body – of his own making, though.

In the writings of the English Cluniac abbot of Gloucester, later bishop of Hereford and London, Gilbert Foliot (d. 1187), we find a proposition which seems to have been composed consciously to promote a literal imitation, but which can still be read easily in an allegorical way. In a letter to one of his friends who had just entered a Cistercian monastery, Gilbert wrote:

> You should follow the crucified One by carrying His cross, express the cross in your habit, contemplate the cross in your mind ... You will carry His stigmata for a bit and fasten His nails into your own flesh. Rough clothing, unprepared food, a hard bed, manual labor, harshness of discipline, [and] continual silence express in you the very nails of the Savior which pierce you. You also

---

10 Constable, *Three Studies*, 202.

11 Loricatus was made famous to post-medieval readers by Edward Gibbon, who in his wonderful prose in *The History of the Decline and Fall of the Roman Empire* (1776-1788) wrote of Loricatus: "By a fantastic arithmetic, a year of penance was taxed at three thousand lashes; and such was the skill and patience of a famous hermit, Saint Dominic of the Iron Cuirass, that in six days he could discharge an entire century, by a whipping of three hundred thousand stripes. His example was followed by many penitents of both sexes; and, as a vicarious sacrifice was accepted, a sturdy disciplinarian might expiate on his own back the sins of his benefactors, ..." (vol. 5, ch. LVIII).

> have the likeness of the lance, which was thrust into His side, in the sting of love that constantly inflicts a wound in your heart.[12]

Often in such texts one encounters a relatively unconcealed desire for Christ himself. This desire, or even craving, is evident in the cult of Jesus, worshipping not only the crucified Christ, but also the man Jesus before the Passion. In these strategies of imitation and performance the absolute negation of the body is no longer prominent.

Quite the contrary: These strategies imply a clear recognition of the corporeal dimension of spirituality. It is exactly the body of the religious athlete which is instrumental in the display of this strategy. With his body the athlete can show the world that he has taken upon himself the sufferings of Christ; the religious idea behind this strategy would have been to relieve Christ himself from His sufferings. Such an *imitatio* strategy could sometimes assume rather conspicuous qualities involving elements of carnal desires and sexual gratification. Reading the examples from the literature, one would even be tempted to characterise such behavior as not solely *imitatio Christi*, but also almost amounting to what could be termed a *penetratio Christi*: While still a young Benedictine in his monastery near Liege, Rupert of Deutz (d. 1129) experienced a vision in which he embraced a wooden crucifix in the church and suddenly felt Christ opening His mouth to receive the deep kiss from Rupert.[13] In one of his many letters, the famed abbot of Cluny, Peter the Venerable (d. 1156), mentions how his mother had embraced a crucifix so hard and intensely that she could not be separated from it even in death.[14] Hugh of Lincoln (d. 1200), the prior of the first English Carthusian house, Witham in Somerset, and later bishop of Lincoln, whose vita was written in the middle of the thirteenth century, was acclaimed for his attitudes towards the people of his community suffering from leprosy,[15] and Hugh would expand his attitudes towards the suffering bodies of the sick into an attitude of reverence of the frail body of Christ as well:

---

12 Constable, *Three Studies*, 212. For an analysis of Gilbert of Foliot and his skill of letter writing see Adrian Morey, *Gilbert of Foliot and his Letters* (Cambridge: Cambridge University Press, 1965).

13 Constable, *The Reformation*, 282, referring to Ruperti Tuitiensis, *De glorie et honore filii hominis super Mattheum*, Corpus Christianorum: Continuatio medievalis 29 (Turnhout: Brepols, 1979), 382-3. For an analysis of the life and work of Rupert of Deutz, see John van Engen, *Rupert of Deutz* (Berkeley: University of California Press, 1983).

14 Constable, *The Reformation*, 281-2, referring to his own edition of Peter's letters: *Peter the Venerable, The Letters of Peter the Venerable*, ed. Giles Constable (Cambridge, MA: Harvard University Press, 1967). The reference goes to Ep. 52, I, 169.

15 Hugh would show great compassion towards those struck with leprosy: "There were hospitals on certain of the episcopal manors, where many men and women afflicted by this disease were maintained. ... He would sit in their midst in a small inner room and would comfort their souls by his kindly words ...Before his address the women withdrew at his command and he went to kiss the men one by one, bending over each of them and giving a longer and more tender embrace to those whom he saw worse marked by the disease." Here the narrator, Adam of Eynsham, breaks of his prose to state to his readers: "Have pity, sweet Jesus, on the unhappy soul of the narrator! I cannot conceal, would that it were concealed from your vengeance, how much I shuddered not merely to touch but even to behold those swollen and livid,

> 'O', he exclaimed, 'how fortunate were the companions of this gracious man! How sweet I should have found it to have beheld and kissed His footprints, or, if it had been possible, to have held close to my heart anything which His hands or any part of His body had touched. What shall I say of His excretions, if it is not impious to call an excretion what flowed form the Tree of Life? Which feelings, I say, should I feel about the sweat, which perchance flowed from the Vessel basin of such great blessedness, owing to His assumption of our infirmity? I would not only carefully collect it, but would devour it with my lips and imbibe it as something sweeter than honey, and would treasure it in the depths of my heart...'[16]

Besides the obvious difficulties in many of these writings in discerning with any sort of precision what is subject to devotion and what is object for devotion, to a modern gaze many of these accounts seem candid and almost erotic. The famed American historian of gender, sexuality, and spirituality, Caroline Walker Bynum, distinguishes between men and women religious in many of her works. For Bynum it is noticeable that the women religious *somatized* their religious experience and gave positive significance to bodily occurrences. These women, to Bynum's mind, seem more immediate in their religious utterances than their male contemporaries. This apparent difference in spirituality need not worry us here. Suffice it to say that women also – and more so, if Bynum are right – speak of tasting God, of kissing Him deeply, or going into His heart or entrails, or being covered by His blood. Their descriptions of themselves or of other women, from a modern point of view, hopelessly blur the line between spiritual or psychological, on the one hand, and bodily and even sexual, on the other. Lidwina of Schiedam and Gertrude of Delft, for example, felt such maternal desire for the Christ-child that milk flowed from their breasts; Beatrice of Nazareth experienced a joy in Christ that contorted her face and wracked her with hysterical laughter; Lukardis of Oberweimar and Margaret of Faenza kissed their spiritual sisters with open mouths and grace flowed from one to the other with an ardor that left both women shaken. The thirteenth-century poet Hadewijch spoke of Christ penetrating her until she lost herself in the ecstasy of love.[17]

---

diseased and deformed faces with the eyes either distorted or hollowed out and the lips eaten away! To an eye darkened by arrogance the pearl of god did not gleam in the mire. But Your servant, whose eyes You had completely blinded to external superficiality, saw clearly the internal splendor, and therefore those seemed to him the more beautiful who outwardly were the most horribly diseased." *The Life of St Hugh of Lincoln I-II*, ed. Decima L. Douie and Dom Hugh Farmer (Toronto and New York: University of Toronto Press, 1961), II, 13-4.

16 *The Life of St Hugh of Lincoln*, II, 13-5.

17 Caroline Walker Bynum, *Fragmentation and Redemption. Essays on Gender and the Human Body in Medieval Religion* (New York: Zone Books, 1991), 190-1.

A traditional Neo-Platonic dichotomy between body and soul thus does not fully explain the mortifications and other obsessions with the body. It is apparent that to the religious athletes the body was considered unworthy and treacherous; the actions of the body were regarded as being in opposition to God. However, at some point, with the new focus on the humanity of Christ from approximately the thirteenth century onwards, the body seemed to earn a new role to play. The body was, in fact, quite necessary for the awareness of sin, the spirit of self-sacrifice and the notion of humility, which were such prominent features in the whole athletics of these religious men and women. The religious athletes not only negated their bodies; they also re-created their bodies into finely calibrated instruments for salvation.

Accordingly, without the treacherous and recalcitrant body there would be no salvation; the body held the possibility for subjugation, discipline, and finally, re-creation. As such, the body itself was both the yardstick for the efforts of the athlete and the medium through which the endeavors of the athlete would be viewed and evaluated, both by the athlete herself and by the on-lookers. Following Caroline Walker Bynum, one might say that in many texts on medieval spirituality what was hitherto believed to be a firm borderline between body and soul suddenly became extremely permeable. The early medieval notion of a rather clear dualism between body and soul was replaced by far more complex interpretations of this relationship the High Middle Ages. The earlier viewpoints separating the two were increasingly replaced by a new ontology of man which included both soul *and* body. The human body seems to have earned a positive meaning in itself during the High Middle Ages.

Allow me to conclude this short article by quoting again from Bynum, who in many of her books has tried to grasp the shifting meanings of body, religion, and gender in the Middle Ages. Bynum works mostly on women religious, but to my mind many of her observations are applicable even across gendered boundaries. She writes in the acclaimed *Holy Feast and Holy Fast*, her seminal work on gender, food, and religion in the Middle Ages:

> Women did not ordinarily speak of these practices as punishment or as escape from the body. Angela of Foligno found the taste of pus 'as sweet as communion;' Beatrice of Nazareth spoke of Jesus' torment as 'healthy wounds;' the author of the nun's book of Unterlinden described the sound of self-flagellation as melody, rising sweet to the ears of the Lord of Hosts. Women's own view of physicality was not, most basically, dualistic. When women spoke of abstinence, of eucharistic ecstasy, of curing and healing through food, they called it *imitatio Christi*. 'Imitation' meant union – fusion –with that ultimate body which is the body of Christ. The goal of religious women was thus to realize the opportunity of physicality. They strove not to eradicate body but to merge their own humiliating and painful flesh with that flesh whose agony, es-

> poused by choice, was salvation. Luxuriating in Christ's physicality, they found there the lifting up – the redemption – of their own.[18]

The bodies of the religious athletes were used as social and religious markers. The body was – in the same way as the religious athletes and their social groupings themselves – marginalised and isolated, but also honored and loved.

[18] Caroline Walker Bynum, *Holy Feast and Holy Fast. The Significance of Food to Medieval Women* (Berkeley and Los Angeles: University of California Press, 1987), 246.

HOLY ISLANDS AND THE OTHERWORLD: PLACES BEYOND WATER

Eldar Heide

Introduction

In this article I attempt to demonstrate that there is a connection between holy islands and notions of an Otherworld beyond water. I believe that the essence of holy islands is their location on the other side of water. One has to cross water to get there and in this respect holy islands are parallel to the Otherworld, which often is placed beyond water, horizontally or vertically. The liminality of certain islands seems to be related to this; they resemble the Otherworld but are located in this world. Thus, they have an intermediate position and are ideal as points of contact with the Otherworld. I also suggest that some islands are "super-liminal," those that are either reachable on foot and thus belong to the mainland in a way, although they are islands, and those that are sometimes submerged or surface only occasionally. I support Holmberg's [Harva's] theory[1] that the main source of the idea that there is an Otherworld underneath us is derived from the experience of seeing an inverted reflection of this world in calm water. To this explanation I add the specification that dry ground is included in the reflection image, which may explain why one (in most accounts) reaches the same dry kind of land no matter if one accesses the Otherworld through gaps in the ground or by jumping into water. I also discuss islands that seem to have their liminality mostly from being off-shore; that is, far away from society, and the relationship between this and Otherworld entrances in the midst of society. My article is an attempt to understand some of the logic behind the major interfaces between this world and the Otherworld, in particular that behind watery interfaces. This does not, however, imply a claim that the whole worldview was a coherent logical system. The evidence gathered for this discussion indicates that going across water – horizontally or vertically – was a more common passage to the Scandinavian gods, especially Óðinn, than has hitherto been realized.

---

1 Uno Holmberg [Harva], "Vänster hand och motsols," *Rig* 1925: 23-36. Uno Holmberg changed his name to Harva in 1927-28 and he is more widely known under the name Harva.

My focus is on Old Norse sources, but I also make use of material from other Scandinavian cultures and neighbouring cultures in Northwestern Europe. I consider this justifiable because most ethnic groups in Northern Europe have been here for at least the last two millennia (and probably far longer) although not in the same area and not with the same population proportionally – and there has been substantial contact between them through the ages. Therefore, we should expect to find many of the same notions among the different peoples like those discussed here, especially the more basic or general ones.[2] The notion of a watery barrier between this world and the Otherworld seems to be more or less universal. I use information from legendary sagas, although they are late, because the motifs in question about going to the Otherworld are so widespread and, therefore, probably ancient. In addition, it is not natural to assume that the motif of going through water to the gods is derived from Christian ideas. Although it is controversial, I also make use of folklore and other late material because it seems that it contains ancient ideas about the topics I discuss. The Icelandic motif of jumping into water in order to go to the land of the fairies is a good illustration of this; motifs found in thirteenth-century sagas and nineteenth-century folklore are close to each other.[3]

I do not intend do discuss water symbolism in general, just the motif of water as a barrier between this world and the Otherworld and questions that derive directly from this. Nor do I intend to discuss holiness or holy places in general, just the essence of holy islands. I will use the term "holy island" in the broad meaning of "island with a special status." The references will refer to eighteenth- and nineteenth-century tradition unless "medieval," "ancient" or the like is stated.

Otherworlds beyond water

What is an island? The obvious answer is: "a piece of land surrounded by water." Our idea of islands is heavily influenced by maps, however, we have a from-above perspective that in pre-modern times was an exception, restricted to mountainous areas. In a cultural context, the concept of islands should be understood from the horizontal perspective of a person standing on a beach looking over to an island. From this perspective an island is a place on the *other side* of water. Most people see it from somewhere else and to them it is basically a place that has to be reached across water. Of course, not only islands are on the other side of water; so is the mainland in many cases.

---

2 Eldar Heide, "Gand, seid og åndevind," Dr. art. thesis (University of Bergen, 2006), 8; pdf-version available at http://eldar-heide.net/.

3 I address the question of late evidence specifically in "More inroads to pre-Christian notions, after all? The potential of late evidence," in *Preprint Papers of The 14th International Saga Conference, Uppsala, 9th-15th August, 2009*; online at http://www.saga.nordiska.uu.se/.

But an island can *only* be reached by crossing water. Therefore I will argue that an island in a cultural context first and foremost is a place on the *other side* of water.

This perspective may be significant for the understanding of holy islands because the *Otherworld* is often placed on the other side of water. In pre-Christian Scandinavia, it appears that most deceased were believed to go to Hel, a cheerless but not painful place which was situated to the north, underground, and beyond the river Gjǫll.[4] The idea that the (or a) realm of the dead is situated beyond a river (often underground) which the dead have to cross is also found throughout Northwestern Europe and in most other parts of the world. It was also common to believe that the departed went to a land beyond the sea (or other large body of water). From Northwestern Europe this idea is attested from the Low Countries (Celtic or Frankish areas) as early as the early sixth century AD. The Byzantine historian Procopius tells that people of this area ferry the souls of the dead to an island off the coast.[5] In Beowulf, the body of King Scyld (Beowulf's father) is laid in a ship that carries him away over the sea.[6] In Scandinavia it seems that one could also go to Hel across the sea. When the god Baldr was dead, his body was placed in his ship, which was set afloat and then set afire, and later we hear of him in Hel.[7] "[I]t is impossible to avoid the impression that Baldr is supposed to go by ship to Hel," Schjødt remarks.[8] A similar boat funeral is described by the tenth-century Arab traveller Ahmad ibn Fadlan, who witnessed it when he visited the Scandinavian Vikings on the Volga.[9] The boat burials that are known archaeologically from Iron Age Northern Europe[10] probably also reflect the idea that the deceased could go to the realm of the dead by boat.

---

4 *Edda Snorra Sturlusonar* (Copenhagen: Gyldendal, 1931), 11, 66; Rudolf Simek (ed.), *Dictionary of Northern Mythology*, trans. Angela Hall [Cambridge: D.S. Brewer, 1993; revised and expanded edition of *Lexikon der germanischen Mythologie* (Stuttgart: Kröner, 1984)], 137; Birgisson Bergsveinn, "Hel. Heimildagildi dróttkvæðra kenninga um fornt hugarfar," cand. philol-verkefni í norrænum fræðum (University of Bergen, 2001).

5 [Procopius of Caesarea], *History of the Wars. Books V and VI*, The Loeb classical library 107 (London: Heinemann and Cambridge, MA: Harvard University Press, 1928), 266-8.

6 R.D. Fulk, et. al. (ed.), *Klaeber's Beowulf and the Fight at Finnsburg*, edition with introduction, commentary, appendices, glossary, and bibliography. Based on the third edition with first and second supplements of *Beowulf and the Fight at Finnsburg*, ed. Fr. Klaeber, Lexington, MA, 1950 (Toronto: University of Toronto Press, 2008), line 26-52.

7 *Edda Snorra Sturlusonar*, 65-66.

8 Jens Peter Schjødt, "The Ship in Old Norse Mythology and Religion," in *The Ship as Symbol in Prehistoric and Medieval Scandinavia. Papers from an International Research Seminar at the Danish National Museum, Copenhagen, 5th-7th May 1994*, ed. Ole Crumlin-Pedersen and Birgitte Munch Thye (Copenhagen: National Museum of Denmark, Department of Archaeology and Early History, 1995), 23.

9 H.M. Smyser, "Ibn Fadlan's Account of the Rus with Some Commentary and Some Allusions to Beowulf," in *Franciplegius. Medieval and Linguistic Studies in Honor of Francis Peabody Magoun*, ed. Jess B. Bessinger jr. and Robert P. Creed (New York: New York Press, 1965), 92-119.

10 E.g., H.R. Ellis Davidson, *The Road to hel. A Study of the Conception of the Dead in Old Norse Literature* (London: Cambridge University Press, 1943), 16-8 and 39-41; Crumlin-Pedersen and Munch Thye (ed.), *The*

Such ideas probably also lie behind grave islets and grave fields beyond water. In northern Norway there are strikingly many Iron Age graves on uninhabitable islets[11] and it is quite common that grave fields from the Iron Age and Viking Age are separated from the settlements by streams.[12] Grave mounds with ditches around them from the same era may be a parallel. This was quite common; the mounds at Borre in Vestfold, Southeastern Norway, are a good example.[13] For at least some periods of the year, such ditches would fill with water and turn the grave-mound into an island, making an island realm of the dead. In some cases archaeological pollen analyses have shown that water plants grew in the ditches.[14] The fact that the water filling would not be permanent in most cases does not count against this reasoning. The ditch could have been there simply to make the mound look higher and bigger and, of course, the dirt in the ditch represented the nearest building material. But if this was the motivation for the ditch, why did the builders not take the rest of this building material and remove the outer edge of the ditch instead of carrying the rest of what they needed from elsewhere? This would even have concealed their trick, but if the ditch was only the casual result of laziness why did the builders in some cases leave four "bridges," path-ways, across the ditch, roughly corresponding to the four cardinal directions?[15] It is hard to escape the conclusion that a ditch proper was intended. This seems to be the point of view of archaeologists, too.[16] The explanation of "grave islands" (of different kinds) has been most clearly pointed out by Nordberg;[17] Old Norse sources and archaeological evidence indicate that there was a blurred distinction between the realm of the dead and the grave. In a way, the grave represented the realm of the dead. If so, the grave beyond water could represent the realm of the dead beyond water.

Ship as Symbol in Prehistoric and Medieval Scandinavia; Anna-Leena Siikala, *Mythic Images and Shamanism. A Perspective on Kalevala Poetry*, FF communications 280 (Helsinki: Suomalainen tiedeakatemia, 2002), 139-41.

11 Børge Evensen, "Gravholmer. Gravlokalisering og samfunn i jernalder i Vesterålen," Hovedfagsavhandling i arkeologi, Institutt for arkeologi (University of Tromsø, 2003).

12 Julie Lund, "Thresholds and Passages: The Meanings of Bridges in the Viking Age and Early Middle Ages," *Viking and Medieval Scandinavia* 1 (2005), 118-9; eadem, "Åsted og vadested. Deponeringer, genstandsbiografier og rumlig strukturering som kilde til vikingetidens kognitive landskaber," Acta humaniora 389, Doktoravhandling i arkeologi (University of Oslo, 2009), 127-9, 256-8.

13 Bjørn Myhre, "Undersøkelser av storhauger på Borre i Vestfold," in *Halvdanshaugen. Arkeologi, historie og naturvitenskap*, ed. Jan Henning Larsen and Perry Rolfsen, Universitetets kulturhistoriske museer, Skrifter 3 (Oslo: Universitetets kulturhistoriske museer, 2004), 203-26; Lund, "Åsted og vadested. Deponeringer," 256-8.

14 Helge Irgens Høeg, "Pollenanalytiske undersøkelser av prøver fra Borre," in *Halvdanshaugen. Arkeologi, historie og naturvitenskap*, ed. Jan Henning Larsen and Perry Rolfsen, Universitetets kulturhistoriske museer, Skrifter 3 (Oslo: Universitetets kulturhistoriske museer, 2004), 235.

15 Myhre, "Undersøkelser av storhauger," 203-26.

16 *Ibidem*, 203; Lund, "Åsted og vadested. 2009," 257.

17 Andreas Nordberg, "Krigarna i Odins sal. Dödsföreställningar och krigarkult i fornnordisk religion," Akademisk avhandling för filosofie doktorseksamen (University of Stockholm, 2003), 73-5.

We also have many examples and indications that the land of the gods – and the realm of the dead connected to it – is located on the other side of the sea. According to Old Norse sources, most men fallen in battle, especially heroes and kings, went to Óðinn's residence, Valhǫll or to Freyja's residence, Folkvangr, of which little is known.[18] Old Norse sources proper do not give explicit information about the journey to Valhǫll,[19] but according to the Danish *Gesta Danorum*, written in Latin by the historian Saxo Grammaticus around 1200, one could ride through the air across the sea to reach Óðinn's residence from the world of humans.[20] This is supported by ibn Fadlan's account. The dead person probably went to Valhǫll because it is referred to as "paradise" and because he was a chieftain – and he apparently went there by boat. The location of Óðinn's residence beyond the sea is also supported by the prose introduction to the Eddic poem *Grímnismál*,[21] which tells of two young brothers who get lost on a fishing trip and are eventually stranded on a shore where they are taken care of by Óðinn and his wife, Frigg, disguised as a smallholder couple. Saxo's account also corresponds to the Eddic poem *Helgakviða Hundingsbana* II (stanza 49),[22] which tells that the dead hero Helgi, in order to reach Valhǫll, has to ride the "bridges of the sky" (*vindhjalms brúar*, probably a variant of the rainbow bridge Bifrǫst in other sources); Valhǫll is situated a very long way to the west of this bridge. Given that Scandinavia is the starting-point, this would place Valhǫll in the western ocean. The motif of crossing the sea to reach Valhǫll also seems to be found in the prose passage *Frá dauða Sinfjǫtla* in the Poetic Edda: The corpse of the hero Sinfjǫtli is ferried across a fjord – and out of the story – by a mysterious ferryman.[23] The end of this journey was probably Valhǫll because Sinfjǫtli was a great hero and therefore should be expected to go there, and the tenth-century poem *Eiríksmál* mentions Sinfjǫtli among the residents of Valhǫll.[24] Schjødt looks upon the ferryman as "a manifestation of Odin, and the incident as such as Odin taking one of his favourites 'home' to Valhalla."[25] An indication that the gods live beyond the sea can also be found in the late legendary *Gautreks saga*.[26] Starkaðr meets Óðinn and the

---

18 *Grímnismál,7*, 14; Sophus Bugge (ed. and comm.), *Sæmundar Edda hins fróða*. (*Norræn Fornkvæði*) (Christiania: Malling, 1867), 77-8.

19 *Grímnismál*, 21 hints that the journey to Valhǫll could involve the crossing of a river (Bugge, *Sæmundar Edda*, 79), but the stanza is hard to interpret. It is discussed by Nordberg, "Krigarna," 223-5.

20 *The History of the Danes* 2 [Saxo Grammaticus: *Gesta Danorum*], trans. Peter Fisher, ed. Hilda Ellis Davidson (Cambridge: Brewer, 1979-80), book 1, 67-9).

21 Bugge, *Sæmundar Edda*, 75-6.

22 *Ibidem*, 200.

23 *Ibidem, 202-3.*

24 Jónsson Finnur, *Den Norsk-islandske skjaldedigtning*, 4 vols. (Copenhagen: Gyldendal, 1912-15; reprint Copenhagen: Rosenkilde og Bagger, 1967-73), vol. B I, 165.

25 Schjødt, "The Ship," 23.

26 *Gautreks saga*, in *Fornaldar sögur Norðurlanda* IV, ed. Guðni Jónsson (Reykjavík: Íslendingasagnaútgáfan, 1954), 28-30.

rest of the gods on an island somewhere in Hordaland, western Norway. It is peculiar that Starkaðr is already on an island when this happens – he has to leave the island where he has been waiting for a favorable wind and row to *another* island to meet the gods – because crossing the sea was a common way of going to the gods? This recalls the account in the *Jómsvíkinga saga* where Earl Hákon goes to an island (*Prímsignd* or *Prímsigð*) in order to get in contact with and sacrifice to his patron goddesses Þorgerðr Hǫlgabrúðr and Irpa.[27]

There also are a couple of examples where a watery barrier separates the giants from the gods and where the humans are placed with the gods; in the Eddic poem *Hárbarðsljóð* the land of giants is separated from the world of humans and gods by a sound that in one place is narrow enough to be crossed by a footbridge.[28] This barrier is a river in the Eddic poem *Vafþrúðnismál* (15-16)[29] and probably in the myth of Þórr's journey to the giant Geirrøðr.[30] The Old Norse myth of the ship Naglfar also implies that one could go by ship from the world of the giants to the world of the gods (and humans?).[31]

Some accounts connect the legendary island of the dead with real islands. The island that Procopius tells of is called Brittia, apparently the island of Great Britain. This tradition has been preserved on the continent to recent times or the island of the dead is one of the islands off Finistère or other real islands in the area.[32] In Welsh tradition there is a connection between the ruler of the island of the blessed (where the chosen go after death) and the Isle of Man.[33] In Old Norse tradition the sea giant (sea god) Ægir/Hlér (both names were used) is associated with the island Læsø, from Hlésey, in the sea of Kattegat,[34] and his wife Rán was the ruler of the realm of the drowned.[35] Gods could also be believed to live on real islands. Tacitus' famous Nerthus passage from the early Roman Age of northern Germany or southern Scandinavia is an example

---

27 See John Megaard, "Hvor sto 'Slaget i Hjǫrungavágr'? Jomsvikingeberetningens stedsnavn og Sæmundr fróði," *alvíssmál* 9 (1999), 29-54. Þorgerðr was probably Hákon's ancestress; Gro Steinsland, *Det hellige bryllup og norrøn kongeideologi. En analyse av hierogami-myten i Skírnismál, Ynglingatal, Háleygjatal og Hyndluljóð*, Doktoravhandling (University of Oslo, 1989; Oslo: Solum, 1991), 24, 122, 220-6, 238.

28 Bugge, *Sæmundar Edda*, 97-9, especially stanza 56.

29 *Ibidem*, 67.

30 *Edda Snorra Sturlusonar*, 106-8.

31 *Gylfaginning*, 37 and *Vǫluspá* 50 (Bugge, *Sæmundar Edda*, 9), *Edda Snorra Sturlusonar*, 71-3; cf. Schjødt, "The Ship," 23.

32 Theo Brown, *The Fate of the Dead. A Study in Folk-eschatology in the West Country after the Reformation* (Ipswich: Brewer for The Folklore Society, 1979), 63-5.

33 Jan de Vries, *Keltische Religion*, Die Religionen der Menschheit 18 (Stuttgart: Kohlhammer, 1961), 87.

34 Simek, *Dictionary of Northern Mythology*, 151; *Edda Snorra Sturlusonar*, 78. *Hlés* in *Hlésey* is the genitive of the name *Hlér*.

35 Helge Rosén, *Om dödsrike och dödsbruk i fornnordisk religion*. Doktorsavhandling, University of Lund (Lund: Gleerup, 1918), 16-8; Simek, *Dictionary of Northern Mythology*, 260.

of this; the goddess lives (in a grove) on a real and apparently inhabited island "in the ocean" (*in insula Oceani*).[36]

The ideas of Valhǫll as the dwelling-place of the chosen dead are closely related to ideas of a pleasant or paradise-like Otherworld beyond the sea, inhabited by fairies or gods. This is well known from classical literature, and Irish literature from the early Middle Ages onwards has many stories of such a land far off in the western ocean, called *Tír na n-Óg*, *Tír fo Thuinn*, *Mag Mell*, and so on.[37] In medieval England the stories of Avalon resemble this.[38] Medieval Scandinavia has indications of the same idea, partly discussed above: There are many indications that Óðinn's residence could be beyond the sea (or a river) and so could the gathering places of the gods and the land of the giants. In Scandinavian late medieval ballads a common motif is the hero obtaining his unique bride from an island or a land beyond the sea, sometimes inhabited by giants *(jǫtnar)* or elves. Probably the first attestation of this is in the sixteenth-century manuscript of the Danish ballad *Ungen Svejdal* (stanza 17-18),[39] which is known in many variants from later times (also *Ungen Svendal*; Swedish: *Hertig Silverdal*, *Unge Svedendal*). It is not unlikely that this motif goes back to a pre-Christian tradition because *Ungen Svejdal* is based upon the Old Norse Eddic poems *Grógaldr* (the first part of the ballad) and *Fjǫlsvinnsmál* (the last part), of which at least the latter in all probability is as old as most other Eddic poems.[40] The ocean-crossing bridal quest is also found in the Saami mythological poem *The Son of the Sun*, which probably has medieval and pre-Christian roots.[41] Saami and Norwegian folklore share a strong tradition where the land of the fairies (Norwegian *hulderfolket* "the hidden people;" North Saami *mearrahálddit* "the sea fairies")

---

36 Tacitus, *Dialogus. Agricola. Germania*, The Loeb classical library 35, edition with translation, *Dialogus de oratoribus* trans. by Sir William Peterson, *Agricola* and *Germania* by Maurice Hutton (London: Heinemann, 1914), ch. 40.

37 de Vries, *Keltische Religion*, 87, 257-9; Jan E. Byberg, "'Dei lukkelege øyane' i norsk folketradisjon. Ei motivgransking," 2 vols., Hovudfagsoppgåve i nordisk (University of Bergen, 1970), 176-8, 200-2; James MacKillop, *Dictionary of Celtic Mythology* (Oxford: Oxford University Press, 1998), 405-6; John Carey, "The Location of the Otherworld in Irish Tradition," in *The Otherworld Voyage in Early Irish Literature. An Anthology of Criticism*, ed. Jonathan M. Wooding (Dublin: Four Courts Press, 2000), 113-9.

38 McKillop, *Dictionary of Celtic Mythology*, 29.

39 Svend Grundtvig, *Danmarks gamle Folkeviser* II. *Naturmytiske Viser. Legendeviser* (Copenhagen: Selskabet til den Danske Literaturs Fremme, 1856), 238-40.

40 The widespread idea that *Grógaldr* and *Fjǫlsvinnsmál* are merely corrupted fragments of a lost **Svipdagsmál* corresponding to the later ballad must be rejected, however; they clearly are independent poems functioning on their own premises [Eldar Heide, "Fjølsvinnsmål. Ei oversett nøkkelkjelde til nordisk mytolog," unpublished master thesis in Old Norse philology (University of Oslo, 1997); pdf-version available at http://eldar-heide.net/. But it is possible that they formed part of a longer, immanent story; cf. Carol J. Clover's theory ["The Long Prose Form," *Arkiv för nordisk filologi* 101 (1986): 10-39] about independent *þættir* [short stories] together forming an immanent saga.

41 Eldar Heide, "Menneskesjela som vind i den samiske folkevisa Solsonen / *Biejjien baernie* / *Beaivvi bárdni*," *Svenska landsmål och svenskt folkliv* 2007: 67-103; pdf-version available at http://eldar-heide.net/. Cf. Harald Gaski, *Biejjien baernie. Beaivvi bárdni. Sámi Son of the Sun* (Kárášjohka: Davvi Girji, 2003).

among other things is a prosperous and happy land beyond the sea, only found through fog or gales.[42] Icelandic and Danish lore have traditions of similar islands found through fog.[43] This fog element is also found in the Irish tradition of *Tír na n-Óg*.[44]

I am trying to show that the question is not island or mainland. The question is: On the other side of water or not. In this perspective the holy island may be seen as a variant of the larger category "supernatural place on the other side of water." Holy islands, however, are still in this world and may be freely visited by mortals; thus they have an intermediate position, are liminal places, ideal as points of contact with the Otherworld. I will return to this point.

Otherworlds under water

First I will address this question: Why is the Otherworld beyond water? One important reason is undoubtedly that death is conceived of as a major barrier and so are rivers, lakes, and the sea. Moreover, death is conceived of as going somewhere else, to a place that we do not know (very well). Accordingly, death implies a journey to a new residence and because of the uncertainty involved it is natural that this journey represents hardships and obstacles, for instance, crossing rivers or the sea. In more general terms one may define the Otherworld as "somewhere beyond our usual range" and this implies obstacles. But some forms of the traditions require an additional explanation: The Otherworld may be *under* water as well.

In Icelandic folklore the world of *álfar* "elves, fairies" is often accessed by jumping into ponds, rivers or the sea.[45] This motif is first attested in the thirteenth-century legendary saga *Þorsteins þáttr bójarmagns*, which apparently borrowed it from a medieval version of the later tales.[46] Similar motifs are also found in folklore elsewhere

---

42 E. g., O. Nicolaissen, *Sagn og eventyr fra Nordland*, 2 vols. (Kristiania: Mallings, 1879-87), I, 19; Reidar Th. Christiansen, "Litt om nordlandsk folketradition," *Håløygminne* 1 (1920): 12-26; Just Qvigstad, *Lappiske eventyr og sagn*. 4 vols., Instituttet for sammenlignende kulturforskning. Serie B: Skrifter. 3, 10, 12, 15, vol. I: *Lappiske eventyr og sagn fra Varanger*, vol. II: *Lappiske eventyr og sagn fra Troms og Finnmark*, vol. III: *Lappiske eventyr og sagn fra Lyngen I*; vol. IV: *Lappiske eventyr og sagn fra Lyngen II og fra Nordland* (Oslo: Aschehoug, 1927-29), II, 433, 439-41, IV, 170-5; Finn Myrvang, *Huh-tetta. Folkeminne ifrå Andøya*, Norsk folkeminnelags skrifter 91 (Oslo: Universitetsforlaget, 1964), 42-4; Byberg, "'Dei lukkelege øyane' i norsk folketradisjon," 38-40, 149, 161-3; Olav Bø, *Trollmakter og godvette. Overnaturlege vesen i norsk folketru* (Oslo: Det Norske Samlaget, 1987), 57.

43 Jón Árnason, *Íslenzkar þjóðsögur og æвintýri. Nýtt safn*, 6 vols. Árni Böðvarsson og Bjarni Vilhjálmsson önnuðust útgáfuna [Reykjavík: Bókaútgáfan þjóðsaga, 1958-61 (1862-64)], II, 300, 318-9, 321, 333, 339, 371, 419; IV, 484, 494, 512, 538, 575, 594, 614, 622.

44 Howard Rollin Patch, *The Other World According to Descriptions in Medieval Literature*. (Cambridge, MA: Harvard University Press, 1950), 45; Byberg, "'Dei lukkelege øyane' i norsk folketradisjon," 179-81, 193.

45 Árnason, *Íslenzkar þjóðsögur*, I, 101-3; III: 162-4.

46 *Saga af Þorsteini Bæarmagni*, in *Fornmanna sögur* 3. Útgefnar ad tilhlutun hins Norræna Fornfræða Fèlags (Kaupmannahöfn, 1827), 176; Heide, "Gand, seid og åndevind," 32, 222-4.

in Northwestern Europe.[47] In Celtic medieval sources and later folklore the land of the dead or the fairies may be under the western ocean and/or under lakes and/or under islands in lakes or the sea.[48] The same is found in Slavic folklore: The realm of the dead is beyond the sea and/or in a wet underworld accessed through a lake.[49] In Beowulf the monster Grendel and his mother live in a moor and when Beowulf intends to kill the mother after he has killed Grendel he dives down into a marsh which is very deep and described as an ocean.[50] In the Icelandic *Grettis saga*[51] a giant lives in a cave which is situated behind a waterfall and can only be accessed through the pool under the waterfall. In the legendary saga *Ketils saga hængs*[52] an ogress wades into the ocean to a *trollaþing*, that is, "thing of trolls."[53]

It also seems that the residences of some Old Norse gods were under the sea or in a marsh: The goddess Sága and Óðinn feast together in *Søkkvabekkr*, that is, "sunken bench," with the waves rippling above them,[54] and Óðinn's wife Frigg resides at *Fensalir*, that is, "bog halls."[55] According to Eddic poems and Snorri, the gods under

47 Hans Siuts, *Jenseitsmotive im deutschen Volksmärchen*, Teutonia. Arbeiten zur germanischen Philologie 19 (Leipzig: Eduard Avenarius, 1911), 50-2, 243-5 (German); R. Hünnerkopf, "Brunnen," in *Handwörterbuch des deutschen Aberglaubens*, ed. E. Hoffmann-Krayer and Hanns Bächtold-Stäubli (Berlin: de Gruyter, 1927), col. 1677 (German); Theo Brown, "West Country Entrances to the Underworld," in *The Journey to the Other World* Mistletoe series, Folklore Society publications 2 (Totowa, N.J.: D. S. Brewer, 1975), 90-103 (English and Welsh); I. Reichborn-Kjennerud, *Vår gamle trolldomsmedisin*, 5 vols., Skrifter utgitt av Det Norske Videnskaps-Akademi i Oslo II, Hist.-filos. klasse, 1927:6, 1933:2, 1940:1, 1943:2, 1947:1 (Oslo: Dybwad, 1928-1947), 1927:6, 56 (Norwegian); Karlis Straubergs, "Zur Jenseitstopographie," *Arv. Nordic Yearbook of Folklore* 13 (1957): 84 (Estonian). A hole in the ice is the entrance; Saami: Just Qvigstad "Lappischer Aberglaube," in *Zur Sprach- und Volkskunde der Norwegischen Lappen*, ed. idem, Oslo etnografiske museums skrifter 1 (Oslo: Oslo etnografiske museum, 1920), 41; Norwegian: Reichborn-Kjennerud, *Vår gamle trolldomsmedisin*. 1927:6, 50.

48 T. Gwynn Jones, *Welsh Folklore and Folk-custom* (London: Methuen, 1930), 52; John Arnott Macculloch, "Celtic Mythology," in *Celtic. Slavic*, ed. Louis Herbert Gray, The Mythology of all Races 3 (Boston: Marshall Jones Company, 1964), 128; MacKillop, *Dictionary of Celtic Mythology*, 359-60, 406. Cf. Byberg, , "'Dei lukkelege øyane' i norsk folketradisjon," 181, 198; Carey, "The Location of the Otherworld in Irish Tradition," 116.

49 W.R.S Ralston, *The Songs of the Russian People as Illustrative of Slavonic Mythology and Russian Social Life* (London: Ellis & Green, 1872), 107-9; Jan Máchal, "Slavic Mythology," in *Celtic. Slavic*, The Mythology of all Races 3, ed. Louis Herbert Gray (Boston: Marshall Jones Company, 1964), 230.

50 Fulk, et. al., *Klaeber's Beowulf*, especially line 1492-4.

51 *Grettis saga Ásmundarsonar*, ed. Guðni Jónsson, Íslenzk fornrit 7, (Reykjavík: Hið íslenzka fornritafélag, 1936), 214-6.

52 *Ketils saga hængs*, ed. Edition Guðni Jónsson, in *Fornaldar sögur Norðurlanda* II (Reykjavík: Íslendingasagnaútgáfan, 1954), 172-3.

53 Cf. Heide, "Gand, seid og åndevind," 211-3, 226-7.

54 *Grímnismál* 7 (Bugge, *Sæmundar Edda*, 77).

55 *Vǫluspá*, 33; *Gylfaginning*, 35 and 49; *Skaldskaparmál* 27, *ibidem*, 6; *Edda Snorra Sturlusonar*, 38, 64, 110.

the leadership of Óðinn gather for drinking parties at the residence of the giant Ægir, who is the husband of Rán, the ruler of the realm of the drowned.[56]

The south Saami used to believe that the ancestors lived in holy mountains called *saajve(aajmoe)*. In the north Saami area this word – in the form *sáiva* – refers to sacrificial sites and to holy lakes, often round, with no rivers flowing into them, believed to be paired with a vertical shaft leading down to a lower one.[57] Probably this is the original meaning because there is broad agreement that *saajve/sáiva* is borrowed from proto-Scandinavian **saiwa-z*, "sea."[58] This interpretation corresponds to the fact that the Saami shaman, the *noaidi*, predominantly used a fish as the vehicle of his soul when he went to the land of the dead[59] and to the fact that going there was referred to as "diving".[60] The *saajve/sáiva* complex suggests that during the time of the proto-Scandinavian language (i.e., the Roman Iron Age) the Scandinavians believed that the ancestors lived in a mountain under water or that this mountain could be reached through water. This could correspond to information from the much later Icelandic *Laxdœla saga*[61]and *Eyrbyggja saga*:[62] Men who drown (Svanr and Þorsteinn) are seen walking into ancestor mountains, which is a realm of the dead that is known from medieval Iceland, similar to

---

56 *Grímnismál* 45, *Lokasenna* (Bugge, *Sæmundar Edda* 85, 113 ff.); Ellis Davidson, *The Road to hel*, 76-7.

57 See K. B. Wiklund, "Saivo. Till frågan om de nordiska beståndsdelarna i lapparnas religion," *Le monde oriental 1916*: 45-74; Louise Bäckman, *Sájva. Föreställningar om hjälp- och skyddsväsen i heliga fjäll bland samerna.* Stockholm studies in comparative religion 13 (Stockholm: Almqvist & Wiksell, 1975); Hans Mebius, *Bissie. Studier i samisk religionshistoria* (Östersund: Jengel, 2003), 82.

58 Cf. Josef Weisweiler, "Seele und See. Ein etymologischer Versuch," *Indo-Germanische Forschungen* 57 (1940): 25-55.

59 Hans Skanke, "Epitomes Historiæ Missionis Lapponicæ. Pars Prima Anlangende de Nordske Lappers Hedendom og Superstitioner" (1728-1731), in *Nordlands og Troms finner i eldre håndskrifter*, ed. O. Solberg (Oslo: Etnografisk museum, 1945), 191-2, 195. Cf. Carl Solander, Carl, "Kort relation, om den danska Missions tilstånd uti Norrige [...]" (1726), in *Källskrifter till lapparnas mytologi*, ed. Edgar Reuterskiöld (Stockholm: Ivar Hæggströms boktryckeri, 1910), 26; Heide, "Gand, seid og åndevind," 136, and Skanke, "Epitomes Historiæ Missionis Lapponicæ," 195; Jens Kildal, "Afguderiets Dempelse og den Sande Lærdoms Fremgang [...]," (1730 and later, in *Nordlands og Troms finner i eldre håndskrifter*, ed. Marie Krekling, Nordnorske samlinger 5, 2 (Oslo: Brøgger, 1945), 140; Sigvard Kildal, "Nogle Observationer om Trommen eller Runebommen [...]," (1730 and later?), in *Det skandinaviske Litteraturselskabs Skrifter* 2, ed. Jan Werner Abrahamson (Copenhagen: Det skandinaviske Literaturselska, 1807), 456 (Abrahamson's title: "Efterretning om Finners og Lappers hedenske Religion"). In Grønvik's reading of the late seventh-century Eggjum rune stone from western Norway (1985) there is (also) "fish transport" to the realm of the dead in the sea: Ottar Grønvik, *Runene på Eggjasteinen. En hedensk gravinnskrift fra slutten av 600-tallet* (Oslo: Universitetsforlaget, 1985).

60 Isaac Olsen, *Om lappernes vildfarelser og overtro. Kildeskrifter til den lappiske mythologi* II (after 1715), ed. Just Qvigstad, Det Kgl. norske Videnskabers Selskabs skrifter 4 (Trondhjem: Det Kgl. norske Videnskabers Selskab, 1910), 45, cf. 46; Heide, "Gand, seid og åndevind," 232-3.

61 *Laxdœla saga*, ed. Einar Ól. Sveinsson, Íslenzk fornrit 5 (Reykjavík: Hið íslenzka fornritafélag, 1934), 46 (chapter 14).

62 *Eyrbyggja saga*, ed. Matthías Þórdarson and Einar Ól. Sveinsson, slenzk fornrit 4 (Reykjavík: Hið íslenzka Fornritafélag, 1935), 19 (ch. 11); cf. Rosén, *Om dödsrike och dödsbruk*, 18.

the *saajve(aajmoe)*.[63] This may seem confusing to us but there are many examples that the same (place in the) Otherworld can be reached through different passages and that one comes to the same dry kind of land even when the passage is watery.

The examples from the *Eyrbyggja saga* and *Laxdǿla saga* also indicate that drowned men did not necessarily come to Rán or that the stay with her was only temporary. The latter seems to be the case in the tenth-century poem *Sonatorrek*, composed by Egill Skallagrímsson in grief for his son, Bǫðvarr, who drowned.[64] Egill says that Rán has taken Bǫðvarr (st. 7-8) but he also blames Óðinn for the drowning (st. 22) and he appears to believe that Bǫðvarr is dwelling with the gods.[65] If so, one could come to the gods by crossing water vertically. Such a motif is also found in two sagas. The clearest example is the fourteenth-century legendary *Egils saga einhenda*:[66] When the hero Ásmundr has lost a sea battle, he jumps overboard and is then stabbed with a spear by Óðinn himself. Ásmundr's ship sinks and it is stressed that no cargo and no wreckage from it ever was found. The stabbing probably means that Ásmundr was chosen by Óðinn to come to his residence to join his warriors[67] and this motif combined with the jumping overboard and the sinking and disappearance of the ship indicates that Ásmundr reached the Otherworld (Óðinn's residence) vertically through water. It fits that this happened near Læsø. Possibly a similar motif appears in *Sǫgubrot af fornkonungum*[68] which is a fragment of the now lost *Skjǫldunga saga* from around 1200. It tells of the Swedish legendary King Ívarr Víðfaðmi, who allegedly conquered much of Northern Europe and who may have died around 700. He was fostered by Hǫrðr, who seems to have been a manifestation of Óðinn.[69] When Ívarr had become an old man, one day before a battle, Hǫrðr insults Ívarr and provokes him to attack him, and Ívarr becomes so furious that he falls overboard. Hǫrðr jumps after him and neither of them surface again. This, too, may be seen as an example of Óðinn taking one of his favourites home to Valhǫll, through water. Perhaps similar notions lie behind lake Odensjön in Scania, southern Sweden, which probably is named after the god Óðinn – if the name is genu-

---

63 For more about ancestor mountains, see: *Eyrbyggja saga*, 9, note 4; Ellis Davidson, *The Road to hel*, 88; Emil Birkeli, *Huskult og hinsidighetstro. Nye studier over fedrekult i Norge*, Skrifter utgitt av Det norske videnskaps-akademi i Oslo II, Hist.-filos. klasse 1943:1 (Oslo: I kommisjon hos Dybwad, 1944), 208.

64 *Egils saga Skalla-Grímssonar*, ed. Sigurður Nordal, Íslenzk fornrit 2 (Reykjavík: Hið íslenzka fornritafélag, 1933), 246-8.ff.

65 Nordberg, "Krigarna," 274-5.

66 *Egils saga einhenda ok Ásmundar berserkjabana*, in *Fornaldar sögur Norðurlanda* III, ed. Guðni Jónsson (Reykjavík: Íslendingasagnaútgáfan, 1954), 365.

67 There are many parallels to this, cf. Simek, *Dictionary of Northern Mythology*, 242.

68 *Sǫgubrot af fornkonungum*, in *Danakonunga sǫgur*, ed. Bjarni Guðnason, Íslenzk fornrit 35 (Reykjavík: Hið íslenzka fornritafélag, 1982), 53-5.

69 Cf. Nordberg, "Krigarna," 135. There are many parallels to this, e.g., the above-mentioned Starkaðr, fostered by Hrossharsgrani = Óðinn.

ine.[70] The lake is situated deep in a crater-like depression in the flat landscape, it has no river flowing into it, and it is believed to be bottomless.[71] This means that it is of the same type as the north Saami *sáiva*, and actually it looks very much like one of them: Lake Pakasaivo in northern Finland. Accordingly, it makes good sense if Odensjön was conceived of as an entrance to Óðinn's residence.

Probably ritual activity connected to springs, wells, ponds, marshes, and so on, too, is evidence of a belief that the Otherworld could be reached through water. The clearest example is sacrifices at such sites, which have been common for thousands of years.[72] Sacrifices were probably deposited in such sites because they were seen as passages to gods/spirits/powers. Rock carvings, too, at least in Northwestern Europe, are connected with the sea, water, springs, marshes, etc.[73] This may be because of the same notion of a watery passage to the Otherworld.

It seems that horizontal and vertical water crossings were conceived of as variants. They often appear in variants of the same or similar stories and usually lead to the same kind of land. In a Saami legend, the sub-sea fairy world is in fact reached by a horizontal boat voyage: A boy goes by boat through fog to the fairy world, and there he can see fish lines with hooks hanging through a fog "ceiling," from human fishermen above. The fish that they catch are sheep in the fairy world, which is a dry one like our

---

70 Stig Isaksson (p. 29; 1958: *Skånes ortnamn. Serie A, bebyggelsenamn. Del 14, Onsjö Härad.* Utgivna av Dialekt- och ortnamnsarkivet i Lund. Lund: Gleerupska universitetsbokhandeln) mentions indices that the name is a late invention, but on the other hand, no other name of it is known.

71 *Nordisk familjebok. Konversationslexikon och realencyklopedi* 12 (Stockholm: Gernandts boktryckeri-aktiebolag, 1888), 101.

72 Berta Stjernquist, "Spring-cults in Scandinavian Prehistory," in *Gifts to the Gods. Proceedings of the Uppsala Symposium 1985*, ed. Tullia Linders and Gullög Nordquist, Boreas 15 (Uppsala: Almqvist & Wiksell, 149-57; eadem, "The Basic Perception of the Religious Activities at Cult-Sites such as Springs, Lakes and Rivers," in *The World-View of Prehistoric Man. Papers presented at a symposium in Lund, 5-7 May 1997, arranged by the Royal Academy of Letters, History and Antiquities along with The Foundation Natur och kultur, publishers*, ed. Lars Larsson and Berta Stjernquist, Kungl. Vitterhets historie och antikvitets akademien, Konferenser 40 (Stockholm: Kungl. Vitterhets historie och antikvitets akademien, 1998), 157-78; Charlotte Fabech, "Samfundsorganisation, religiøse ceremonier og regional variation," in *Samfundsorganisation og Regional Variation. Norden i romersk jernalder og folkevandringstid. Beretning fra 1. nordiske jernaldersymposium på Sandbjerg Slot 11.-15. april 1989*, ed. Charlotte Fabech and Jytte Ringtved, Jysk Arkæologisk Selskabs skrifter 27 (Århus: Jysk Arkæologisk Selskab), 283-303; Torsten Capelle, "Quell- und Brunnenopfer," in *Opferplatz und Heiligtum. Kult der Vorzeit in Norddeutschland*, ed. Ralf Busch, Veröffentlichung des Helms-Museums 86 (Neumünster: Wachholtz, 2000), 101-5; idem, "Mooropfer," in *ibidem*, 123-7; Stefan Brink, "Mythologizing Landscape. Place and Space of Cult and Myth," in *Kontinuitäten und Brüche in der Religionsgeschichte. Festschrift für Anders Hultgård zu seinem 65. Geburtstag am 23.12.2001*, Ergänzungsbände zum Reallexikon der Germanischen Altertumskunde 31 (Berlin: de Gruyter, 2001), 96-8.

73 Knut Helskog, "The Shore Connection. Cognitive Landscape and Communication with Rock Carvings in Northernmost Europe," *Norwegian Archaeological Review* 32 (1999), 73-94; Lasse Bengtsson, *Bilder vid vatten. Kring hällristningar i Askum sn, Bohuslän*, GOTARC. Serie C, Arkeologiska skrifter 51 (Stockholm: Göteborgs universitet. Institutionen för arkeologi, ANL, 2004).

own.[74] The fact that fog is quite common in connection with vertical water crossings also indicates that the horizontal water crossings and the vertical ones are variants. Saxo Grammaticus (around 1200) tells that the hero, Haddingus, was once led to an underground Otherworld that resembles several of the Old Norse forms of Otherworld. On his way down, he passed through fog (and then crossed a river).[75] A similar motif is attested in the thirteenth-century *Þorsteins þáttr bǿjarmagns*. In order to go to the land of the *álfar* the hero jumps into a river and then, on the way down, passes through fog.[76] The same motif is found in a rich early nineteenth-century Icelandic tradition about going to the *álfar*.[77] In the horizontal water crossings to the Otherworld, fog is a necessary element because it hides the location of the Otherworld (is)land. This hiding function of fog is sometimes found in connection with journeys on land as well.[78] In the vertical variant, however, the motif has no apparent function. The only possible explanation of it seems to be that it is transferred from the horizontal variant – which means that the vertical and horizontal water crossings were conceived of as variants.

The different Otherworlds

At this point one might object that I oversimplify the notions of the Otherworld, that I ignore the differences between the various forms, but my claim is not that all forms are the same. There are many major differences. For instance, the lands of the fairies, elves, gods, etc., in folk traditions are not usually conceived of as realms of the dead, or at least not first and foremost. But for my purpose this does not matter because I simply discuss the watery barrier between this world and Otherworlds, no matter what Otherworlds we are talking about. Most traditions do not make a sharp distinction between the realm of the dead and the other Otherworlds. This perhaps applies most clearly to Irish tradition.[79] There, the land of the fairies, the *Tír na n-Óg* and so forth, and the realm of the dead are frequently located in the same places and described in the same ways. Moreover, after death the chosen may reappear among the fairies or in the blessed land beyond the sea. The latter also applies partly to Old Norse tradition, as we have seen; the realm of the gods is not primarily a realm of the dead, but after

---

74 Christiansen, "Litt om nordlandsk folketradition," 19.

75 Fisher and Ellis Davidson, *The History of the Danes* 2, book 1, 8,14.

76 *Saga af Þorsteini Bæarmagni*, 176.

77 Árnason, *Íslenzkar þjóðsögur* I, 101-3; III, 162-4.

78 Icelandic tradition: Einarsson Bjarni, *Munnmælasögur 17. aldar*, Islenzk rit síðari alda 6 (Reykjavík: Hið Íslenzka fræðafélag, 1955); Irish tradition: Patch, *The Other World*, 45; Macculloch 1964, "Celtic Mythology," 118.

79 Seán Ó Súilleabháin, *A Handbook of Irish Folklore* (Dublin: The Folklore of Ireland Society, 1942), 450; Arthur C.L. Brown, *The Origin of the Grail Legend* (Cambridge MA: Harvard University Press, 1943), 23; MacKillop, *Dictionary of Celtic Mythology*, 359-60.

death heroes and men killed in battle come to Óðinn's or Freyja's residence. Among the (south) Saami the people of the *saajve(aajmoe)* were believed to be ancestors, but the description of them and their residence makes it hard to distinguish them from elves or fairies.[80] The same applies to the ancestor mountains in medieval Iceland. Holmberg [Harva] points out that the fairies (*de underjordiska*, that is, "the 'subterraneans'") in Finnish tradition are "very closely related to the spirits of the dead" and that in many cases these groups are identical in the traditions of the Finno-Ugric peoples of Russia.[81]

Reflection in water and interfaces with the (inverted) Otherworld

How can one explain the vertical water crossing to the Otherworld? This question is part of the bigger question of why the most common notion of Otherworld places is *underneath* us. The underneath part is often explained by the fact that most cultures bury their dead in the ground, which easily leads to a notion that there is an Otherworld in the ground, but it does not explain the vertical crossing of *water* to get there. Nor does it, as Holmberg [Harva] points out,[82] explain why even peoples who do not bury their dead believe in an Otherworld below us – such as the Siberian Evenks, who place their dead high in the air on platforms and consider it a sin to bury them in the ground. Holmberg [Harva] also remarks that the underground Otherworld of folklore and myth is not usually situated at the depth of a grave. He launches another explanation of the Otherworld beneath us.[83] He points out that in many cultures the Otherworld is conceived of as *inverted* from this world; it is upside-down, the rivers flow backwards, the people walk backwards, are left-handed, wear their clothes inside-out, etc. Holmberg has information of such notions from most of northern Eurasia. From some areas he even has information that the sun was believed to move from right to left, and in some accounts, from Sweden (Gotland), Saami land and Nenets land (northwest Russia), it is even said that the "Otherworlders" walk with their feet against ours. Holmberg points out that a world like this seems to be derived from the reflection in lakes and ponds on calm days. Not only is it upside-down; it is also inverted left/right.[84]

---

80 Skanke, "Epitomes Historiæ Missionis Lapponicæ," 190-2; Wiklund, "Saivo," 55; Bäckman, *Sájva*, 67-9; and Mebius, *Bissie*, 81-3.

81 Holmberg [Harva], "Vänster hand och motsols," 25-6.

82 *Ibidem*, 33.

83 *Ibidem*, 33-5; idem, *Finno-ugric, Siberian [Mythology]*, The Mythology of all Races IV, ed. John Arnott et al. (Boston: Marshall Jones Company, 1927), 73.

84 A variant is the notion that there are night and winter in the otherworld when there are day and summer in this world, although this cannot be directly derived from the reflection in the water, just from the general inversion of the mirror image. Saxo's account of the underworld visited by Haddingus is an early

Holmberg [Harva] based his work on theories of mental evolution that are now rejected, but the experience of an inverted world that can be seen in a reflection in water is shared by all seeing humans through all ages, regardless of cultures and theories. Accordingly, the explanation in question is not dependent upon evolutionist theories and it was put forward by Carl von Linné in 1745 (as mentioned by Holmberg), long before the invention of evolutionism.[85] To me it seems hard to escape the opinion that Holmberg [Harva] was right: The main source for the notion of an Otherworld underneath us is the reflection in water. This notion can be as old as mankind and yet renewed again and again as new generations have seen reflections.

In particular, the reflection can explain the notion of an Otherworld underneath us beyond water. It may also explain much of the general idea of an Otherworld beyond water, even in a horizontal direction, because vertical and horizontal water crossings seem to be variants in the traditions. The experience of (horizontal) crossings of rivers, lakes, and seas on journeys is part of the background, too, connected to the notion of death as a (horizontal) journey to a distant land.

If one accepts the reflection-in-the-water explanation, one would expect water to be the most common interface with the Otherworld. Whether this is so or not is hard to find out, but at least it is common throughout the world (as has been noted by many scholars[86]); the Otherworld is accessed by jumping into a well, pond, lake or the sea, or a hole in the ice, or the spirits or beings from the Otherworld come up to this world through the same kinds of watery passages. Equally common entrances to the Otherworld, however, are gaps in the ground, caves, holes in the ground, clefts, volcanoes[87] and door-like formations in rocks,[88] or grave entrances.[89] These types usually exist side by side with no apparent distinction, which may seem contradictory but in my opinion is not. A well is a gap in the ground, too, just filled with water. A large lake or the sea are different but they, too, are potential gaps. They form a gap as soon as one jumps

---

Scandinavian attestation of this; it has summer and fresh plants when there is winter above the ground (Fisher and Ellis Davidson, *The History of the Danes* 2, book I, 30-1).

85 Carl von Linné, Carl Linnæi öländska och gothländska resa på riksens höglovlige ständers befallning förrättad åhr 1741. Med anmärkningar uti oeconomien, naturalhistorien, antiquiteter & c. med åtskilliga figurer (Stockholm: Gottfried Kiesewetter, 1745), 312-3.

86 E.g., Brink, "Mythologizing Landscape," 96.

87 For instance: Jones, *Welsh Folklore*, 52 (Wales); MacKillop, *Dictionary of Celtic Mythology*, 359 (Ireland); Brown, "West Country Entrances to the Underworld;" (England and Wales); Árnason, *Íslenzkar þjóðsögur*, I, 101-3; III, 162-4 (Iceland); Feilberg 1910: 59-60, Myrvang, *Huh-tetta*, 29-30, 35, 38 (mainland Scandinavia); Bäckman, *Sájva*, 126 (Saami land); Straubergs, "Zur Jenseitstopographie," 83-4 (Estonia).

88 Ernst Manker, *Lapparnas heliga ställen. Kultplatser och offerkult i belysning av Nordiska museets och Landsantikvariernas fältundersökningar*, Acta Lapponica XIII (Stockholm: Gebers, 1957), 195-7, image 172-4 (Saami land).

89 *Ibidem*, 162. Cf. Straubergs, "Zur Jenseitstopographie," 84 (Estonia); Nordberg, "Krigarna," 74-6, 273-5 (medieval Scandinavia).

into them and sometimes the passage to the Otherworld through water is formed by a whirlpool,[90] which is a gap, even a shaft, in the water. More important, in a small calm puddle surrounded by high trees or mountains one can "literally" see through a gap and deep into the Otherworld (even if the puddle is no more than a centimetre deep). If one looks into a lake the gap part is not there but still one can get the impression of looking into the Otherworld; it appears that one usually comes to the same kind of land no matter if the entrance is watery or not. In most accounts the sub-water Otherworld is the same dry kind of land that we live on; it is not a land of fish, seaweed and mermaids, which is why I often use unusual expressions like "Otherworld vertically beyond water" instead of simply "under water." The latter gives the impression of a watery world. If one looks at the reflection in a pond this is natural: In the water one sees dry land, exactly like the land that we walk on. The Otherworld accessed through water is just that: Accessed through water. It is not necessarily a watery world; even when the Otherworld rises from the sea as an island a dry type of land is revealed. This reduces the difference between the two probable (main) starting-points of the notion of the underworld: The reflection in water and graves in the ground. The reflection in water includes the ground, so to speak.

It is worth noting that there was an Otherworld for the gods, too, not only for humans, and that the gods accessed the Otherworld through the same kinds of entrances as humans. When the Old Norse god Baldr was killed he went to the same realm of the dead as most humans, Hel, by boat. When the god Hermóðr went to Hel to rescue Baldr he experienced the same kind of journey as humans, downwards and across a river. Þórr met giantesses on the island Læsø, which seems to be a place where humans can get in contact with the Otherworld (see below). Óðinn seeks contact with supernatural powers in a well,[91] and so on. I see no contradiction in this.

"Super-liminal" islands

One may in any way accept or reject the reflection-in-water explanation of the Otherworld below us, but it does not seem to matter if the Otherworld is beyond water horizontally or vertically. It seems that the barrier or border aspect of the water is the same in both cases and that this aspect is what matters. This can throw light upon some special cases of islands. As already noted, islands in general are natural as places of contact with the Otherworld because in a way they are situated "on the other side," al-

---

90 Holmberg [Harva], *Finno-ugric, Siberian [Mythology]*, 78-9; Veikko Anttonen, "The Concept of *pyhä* (Sacred) in Pre-Christian Finnish Religion," In *Northern Religions and Shamanism*, ed. Mihály Hoppál and Juha Pentikäinen, Ethnologica uralica 3 (Budapest: Akadémiai Kiadó, 1992) 36; Nordberg, "Krigarna," 39. Cf. "sucked down" in some German accounts: Siuts, *Jenseitsmotive im deutschen Volksmärchen*, 54-5.

91 *Vǫluspá* 27-8 (Bugge, *Sæmundar Edda,* 9).

though they are in this world. But some islands seem to have an increased island liminality.

In some cases this seems to be due to a combination with other principles. Some islands have entrances to the Otherworld on them, in caves, clefts or the like. An example from south Saami land is the island of *Tjåehkere* (South Saami) or *Gudfjelløya* (Norwegian) in Lake Tunnsjøen in Trøndelag (central Norway).[92] The island is a mountain and on top of it there is a narrow deep cleft that splits the mountain in two and has been believed to reach all the way down to the lake, 400 metres below.[93] The island is sacred in the tradition of the Saami, who used to sacrifice reindeer to the island's god in the cleft. Another example is the small island of *Äjjih* (Inari Saami)/*Ukonsaari* (Finnish) in Lake Inari, Finnish Lapland,[94] which has a sacrificial cave.[95] In these cases I do not know of traditions that the cleft/cave was an entrance to the Otherworld, but this seems a reasonable assumption because of the sacrifices deposited there. There are other such traditions in Estonia[96] and Ireland:[97] In the sea/lake/marsh there is an island; on the island there is a cave/hole, and this is an/the entrance to the Otherworld. The same motif may be seen in the fourteenth-century Icelandic *Bárðar saga*,[98] although in this case the Otherworld is only the dwelling place of a deceased evil king and his men. Similar notions may have been connected to the islands in Salten, northern Norway, that were called *Goðeyjar* ("islands of the gods") in Old Norse. There are several large grave mounds on them and the place was one of northern Norway's major heathen strongholds in the age of conversion.[99] The name *Goðeyjar* is probably related to Saltstraumen, the world's strongest tidal current, which flows between the islands. But why exactly were these tidal-current islands dedicated to the gods and such a powerful site of pagan worship? It may have been because they were associated with the whirlpools that gave access to the Otherworld. The island part may also be significant, how-

92 The Norwegian name means "the island of the god mountain" and the Saami name probably refers to the thunder god.

93 Agnar Nordback, "Tunnsjø og Tunnsjøguden," *Årbok for Nord-Trøndelag Historielag* 1933: 63-66; Martin Bjerken, *Grong bygdebok. For Grong, Harran, Namsskogan og Røyrvik* (Trondheim: [s.n.], 1950).

94 The names of this island, too, refer to the thunder god. Thanks to Håkan Rydving, University of Bergen, for help with the names *Tjåehkere* and *Äjjih*.

95 Manker, *Lapparnas heliga ställen*, 292, image 338-40.

96 Straubergs, "Zur Jenseitstopographie," 84.

97 MacKillop, *Dictionary of Celtic Mythology*, 406.

98 *Bárðar saga [Snæfellsáss]*, in *Harðar saga*, ed. Bjarni Vilhjálmsson and Þórhallur Vilmundarson, Íslenzk fornrit 13 (Reykjavík: Hið Íslenzka fornritafélag, 1991), 165-7.

99 *Óláfs saga Tryggvasonar*, in *Heimskringla* I by Snorri Sturluson, ed. Bjarni Aðalbjarnarson, Íslenzk fornrit 26 (Reykjavík: Hið íslenzka fornritafélag, 1941), 324-6, *Rǫgnvalds þáttr ok Rauðs*, Richard Perkins, *Thor the Wind-Raiser and the Eyrarland Image*, Viking Society for Northern Research text series XV (London: Viking Society for Northern research, 2001).

ever, in this case as well as in other cases. An entrance to the Otherworld makes a liminal place and if it is associated with an island the liminality may be doubled.

Another type of island seems to have a special status derived from a highlighting of the island liminality. There are two types of such "super-liminal" islands; one can be reached on foot and one is sometimes under water, sometimes over. The first type is commented on by Manker in his book on the holy (sacrificial) places of the Saami.[100] He remarks that surprisingly many of the holy islets are separated from the mainland by sounds so shallow that they can be forded. I would like to add that some of them have a sub-water ridge leading to the islet. One might think that such islets were chosen because access was easier if a boat was not required, but it would have been even easier to choose some site on the mainland and there is explicit information (from 1727 and later) that difficult-of-access places were preferred.[101] An explanation may be that islets/islands that can be reached on foot are "exceptionally in-between." Being surrounded by water they are islands, but being reachable on foot they belong to the mainland; therefore they may have been seen as being situated on the very border between worlds and therefore especially suitable as points of contact with the Otherworld. – All of Manker's examples are from lakes. On the coasts of Scandinavia the shoreline has changed a great deal since pre-Christian times because of the post-glacial rebound and therefore islands like this are hard to identify in most areas. Heligön, "the holy island" on the southeastern coast of Gotland, is a possible example, however, because there post-glacial rebound has been virtually nil. The sound that separates Heligön from the mainland is "bridged" by a shallow sub-water sand ridge (a *val* in Norwegian, from Old Norse *vaðill*, m.). Possibly such a sub-water ridge also connected Helgö "the holy island" and the mainland in Lake Helgasjön in Småland, southern Sweden. Hyltén-Cavallius refers to it as "a stone bridge or a ford" called "the Holy Bridge."[102] The words are confusing; either it is a bridge or it is a ford. Possibly the name made Hyltén-Cavallius use the term bridge although it was a ford, a sub-water ridge. There may be another example at Helgøya, "the holy island" (Old Norse *Eyin Helga*), in Lake Mjøsa, eastern Norway, the country's largest lake. The pre-Christian cultic centre of the island was probably the farm *Hofvin* (Old Norwegian; now *Hovinsholm*); the name means "the field by the temple." This farm is situated on the southern tip of the island, facing an islet (called Holmen, "the islet") which seems to have enjoyed a special status in pre-Christian times; it has at least one large grave mound (35 m in diameter, 6 metres high) and some rampart construction. It is separated from Helgøya by a sound that is often fordable in spring.

---

100 Mankers, *Lapparnas heliga ställen*, 24.

101 *Ibidem*, 28.

102 Gunnar Olof, Hyltén-Cavallius, *Wärend och Wirdarne. Ett försök i Svensk Ethnologi* 2 (Stockholm: Norstedt, 1863-68), I, 130: "en stenbro eller ett vad."

Many of these shallow-sound islands may be grouped in a sub-type: Islands that can *sometimes* be reached on foot – in certain seasons (Holmen near Helgøya, Helgö in Småland?), or during anticyclones (Heligön near Gotland?), or at low tide. (The Baltic Sea does not have tides but fluctuates a great deal according to the air pressure and winds.) An example of an island that can be reached at low tide occurs in *Bárðar saga*.[103] The passage tells how the hero, Gestr, and his men are sent from Trondheim, Norway, to kill a monster ghost that lives in a grave mound in Helluland, which is probably Baffin Land in arctic Canada. When they finally get there it turns out that the grave mound is situated on an islet that is connected to the mainland by a ridge (*rif*) which is submerged except at low tide. Gestr and his men dig the mound open and Gestr is lowered into the mound on a rope through a shaft that is revealed. There he kills the monster ghost. When Gestr and his men have completed their mission they cannot go back to the mainland because the ridge is submerged by the raging sea – until the priest separates the sea and rescues the men, like many times before in the story. It seems that the location of this entrance to the Otherworld is liminal in two ways: It is situated in the remotest possible place in the known world and it is situated on a "semi-islet" that only belongs to the mainland part of the time. The fact that the raging sea that blocks contact with the mainland apparently represents the Otherworld powers seems to confirm this understanding: The islet is on the very border between the worlds, shifting back and forth between this side and the other side – not only topographically but also in terms of who has the power. One might object that the source value of this saga is low because it is late, fantastic, and heavily influenced by Christianity – the priest and the holy King Óláfr play important roles. The motif in question, however, is not Christian, so it is probably pre-Christian although it is used in a Christian understanding.[104] It is possible that the grave mounds with ditches around them should also be understood as being "exceptionally in-between" in this way (see above). Many of the ditches would probably fill with water during heavy rain and/or the snow-melt in spring, but few would do so permanently. Thus, they may have been seen as symbolically shifting back and forth between worlds.

The other main type of "super-liminal" islands is the one that is completely submerged for periods of time. This is the case with many Saami inland holy "islets" (sacrificial sites) that are just boulders; they are so low that they are submerged during flooding. The special status of these places also seems to derive from a highlighting of the island liminality, but in a different way. All islands are in a way "on the other side," but these islands are even more so because they are sometimes submerged and then belong to the Otherworld completely. They are still accessible, however, in the periods when they are not submerged. This may be compared to the fact that rock carvings

---

103 *Bárðar saga*, 165-7.

104 Heide, "More inroads to pre-Christian notions, after all?"

often are placed where running water, waves or the high tide cover them.[105] In the case of the sea this has been established by analogy with lakes, Onega in particular, where the shores have no post-glacial rebound. The Saami boulder islets were often believed to be dwelling-places of the *háldit*, "fairies, elves" (North Saami), who lived on such "islets."[106] Accordingly, they are related to the sea islands inhabited by fairies – *hulderfolket*, *mearra-háldit* – in Norwegian and Saami folklore. In Norwegian tradition the most famous are Utrøst and Sandflesa.[107] Most, like these two, are offshore, but some are situated among the coastal islands and some even in fjords.[108] They are usually submerged but occasionally surface and may then be seen or may be found by fishermen in fog or gales; a few of them are actual rocks but most of them only exist in popular lore. Some forms of the Irish traditions of *Tír na n-Óg*, and so on, are similar to the Norwegian traditions. In these traditions we again see that this world and the Otherworld meet on islands whose liminality is highlighted by the water surface shifting between this side and the other side. The introduction to the thirteenth-century *Guta saga* seems to be an early example of this.[109]

Remote islands

The Utrøst-type of island is related to a type which is not periodically submerged but which has a similar location, that is, remote in the geography of this world. The islands of Læsø (Old Norse Hlésey) between Jutland and Sweden and Samsø (Old Norse Sámsey) between Jutland and Zealand are good examples of this. They have been moderately important economically, but have a prominent position in Old Norse poetry[110] and the legendary sagas.[111] (Samsø was important strategically, however.) According to the Eddic poems *Lokasenna* and *Hárbarðsljóð*, these islands received visits from the gods Þórr and Óðinn[112] and the giant Ægir/Hlér was associated with the island. This is special because the mythological world had its own geography; gods and giants did not

---

105 Helskog, "The Shore Connection;" Bengtsson, *Bilder vid vatten.*

106 Wiklund, "Saivo," 50.

107 Byberg, "'Dei lukkelege øyane' i norsk folketradisjon."

108 *Ibidem*, 86-88.

109 *Guta saga. The History of the Gotlanders*, ed. Christine Peel, Viking Society for Northern Research, text series XII (London: Viking Society, University College, 1999), 2.

110 Finnur Jónsson, *Lexicon poeticum* (Copenhagen: S. L. Møllers Bogtrykkeri, 1913-16), 261, 484.

111 *Hervarar saga ok Heiðreks*, ed. Guðni Jónsson, in *Fornaldar sögur Norðurlanda* II (Reykjavík: Íslendingasagnaútgáfan, 1954), 1-71; *Örvar-Odds saga*, ed. Guðni Jónsson, in *Fornaldar sögur Norðurlanda* II (Reykjavík: Íslendingasagnaútgáfan, 1954), 199-363; *Göngu-Hrólfs saga*, ed. Guðni Jónsson, in *Fornaldar sögur Norðurlanda* III (Reykjavík: Íslendingasagnaútgáfan, 1954), 167; *Egils saga einhenda*, 365; *Ragnars saga loðbrókar*, ed. Guðni Jónsson, in *Fornaldar sögur Norðurlanda* I (Reykjavík: Íslendingasagnaútgáfan, 1954), 284-5.

112 *Lokasenna*, 24, *Hárbarðsljóð*, 37 (Bugge, *Sæmundar Edda*, 117, 101).

normally wander around in this world, but apparently on Læsø and Samsø they did. This indicates that these islands were places where this world and the mythological world could meet. It seems that Snorri conceived of Læsø in this way when he wrote the framing story for *Skaldskaparmál*, part of his *Younger Edda*. There he lets gods, giants, and elves meet for a party on Læsø.[113] Two legendary sagas also give the impression that Læsø was a point where the different worlds could meet. In *Gǫngu-Hrólfs saga* there is a fantastic account of the berserk Grímr, who grew up on Læsø;[114] his mother was believed to be a sea ogress (*sjógýgr*) and he was found (as a baby) floating in the sea by the high-water mark – which is the boundary between the worlds? The sea-battle mentioned in *Egils saga einhenda* is another example that Óðinn could enter this world on Læsø.[115] In the legendary saga *Hervarar saga ok Heiðreks* (1954:13) a different kind of Otherworld contact takes place on Samsø:[116] A *haugbrot*, "grave-mound robbery." There are many accounts of *haugbrot*s in the Old Norse sources, especially in the legendary sagas,[117] but the one in the *Hervarar saga* is unique because the hero – a heroine – cooperates with the dead in the mound, who are her ancestors, and because of this the mound opens willingly without her digging it open. Thus, this motif is reminiscent of the Helgafell motif in the *Eyrbyggja saga*. According to the legendary *Ragnars saga loðbrókar* there was a cult site with a wooden idol on Samsø in pagan times.[118] The motif that a remote island is a place of contact with the Otherworld is also found in the legendary *Ketils saga hǫngs*:[119] The hero is alone, fishing off-season on the island of Skrova in northern Norway, which has a location similar to Læsø and Samsø, isolated, between the mainland and the Lofoten archipelago, but quite far from both. Here he meets an ogress who tells about events in the world of trolls. Essentially the same motif is common in Norwegian and Saami folklore: The hero meets the fairies when visiting out-of-season an isolated island used for seasonal fishing.[120] The motif is found in Icelandic folklore as well, attributed to the island Seley off the eastern coast of Iceland, near Reyðarfjörður.[121] These examples resemble those given above, where contact with gods and the like is sought on

---

113 *Edda Snorra Sturlusonar*, 78; cf. *Grímnismál*, 45 and *Lokasenna* (Bugge, *Sæmundar Edda*, 85, 113-5).

114 *Göngu-Hrólfs saga*, 167.

115 *Egils saga einhenda*, 365.

116 *Hervarar saga ok Heiðreks*, 13.

117 A.W. Brøgger, "Oseberggraven – Haugbrottet," *Viking* 1944: 1-45; Heinrich Beck, "*Haugbrot* im Altnordischen," in *Zum Grabfrevel in vor- und frühgeschichtlicher Zeit. Untersuchungen zu Grabraub und "haugbrot" in Mittel- und Nordeuropa*, ed. Herbert Jankuhn et. al., Abhandlungen der Akademie der Wissenschaften in Göttingen, philologisch-historische Klasse, 3. Folge (Göttingen: Akademie der Wissenschaften, 1978), 211-28; Jonas Wellendorf, "Haugbrot i norrøn litteratur og omegn," hovedoppgave i nordisk (University of Bergen, 2002).

118 *Ragnars saga loðbrókar*, 284-5.

119 *Ketils saga hængs*, 172-73; cf. Heide, "Gand, seid og åndevind," 211-3, 226-8.

120 Byberg, "'Dei lukkelege øyane' i norsk folketradisjon," 58-60.

121 Árnason, *Íslenzkar þjóðsögur* I, 86-8.

islands, although in those cases it is not always a point that the island or the place of contact on it is remote.

This type of island seems to have its liminality highlighted by a combination of the island motif and a different liminality motif: Remoteness. Læsø is an island far off-shore and so is the Irish *Tír na n-Óg* in most accounts. Most of the Norwegian fairy islands are off shore as well and many of them have names with the prefixes *Ut-*, *Ytter-* or *Aud-* "Outer" and "Deserted" (Ut-Røst, Ytter-Sklinna, Audvær, etc.).[122] Samsø is not strictly off-shore, but relatively seen it is, being a small island quite far away from all the surrounding larger islands and Jutland.

The motif of a man meeting the fairies (or other Otherworld beings) when visiting a remote island off season also applies to summer farms in the mountains (at least in Scandinavia).[123] How can one understand this parallelism? This question is part of the larger question of how we can understand the liminality that derives from remoteness. We have seen repeatedly that the Otherworld can be reached by going a long way horizontally or downwards through a nearby entrance.[124] These may seem like completely different ways, but this is not necessarily so. The Otherworld is beyond the areas that we usually (can) access. Our horizontal range is long so therefore we have to go a long way horizontally to reach the Otherworld, especially if the distance alone separates us from the Otherworld. Downwards, however, our range is very limited. Therefore, even when the downward entrances to the Otherworld are next to where we live, for instance, in the middle of the farmyard (a well or a grave mound), the Otherworld is very far away. In both cases, however, the essence is going beyond what we usually (can) access. Both the seasonal fishing islands and the summer farms are outposts; they represent the outer edges of the society. Because the Otherworld can be defined as "the world beyond (known) human society" such outposts have a liminal location; they are on the border. The Finnish places designated as *pyhä* "sacred" in pre-Christian times were always boundary markers, among other things.[125] They even change sides through the year when they are abandoned by the humans and occupied by the fairies or other Otherworld powers at the end of the season and then reoccupied by the humans next year. They are periodically "submerged," so to speak, like one of the types of "super-liminal" islands.

---

122 Byberg, "'Dei lukkelege øyane' i norsk folketradisjon," 86-8.

123 E.g., P. Chr. Asbjørnsen, *Norske Huldre-Eventyr og Folkesagn.* 3rd ed. Christiania: Steensballe, 1870); Svale Solheim, *Norsk sætertradisjon*, Instituttet for sammenlignende kulturforskning, Serie B, Skrifter 47 (Oslo: Aschehoug, 1952), 371-3, 376-8.

124 In addition, the otherworld can be reached upwards, by going to the top of a high mountain or going to the sky or stars. This is a motif I have not discussed in this article but will return to in the concluding remarks.

125 Cf. Anttonen, "The Concept of *pyhä*," 13.

Islands like Læsø, Skrova and other seasonal fishing islands are liminal both as islands and as far-away places. On top of this, these islands are only seasonally inhabited, shifting sides through the year. This logic probably lies behind islands like Helgö(n) "the holy island" (now mostly called Storjungfrun) off the coast of Hälsingland in northern Sweden. Brink points out that the island is far from the coast and suggests that it was a place of contact with the gods and/or a home of dead relatives. [126] This seems plausible to me, and I would like to add that Helgö(n) in older times was inhabited only seasonally, by fishermen. As far as the ancestor part is concerned, Brink's explanation would make the island parallel to islands like those off Finistère, the Isle of Man, the island of Great Britain, and Læsø, which have been conceived of as (or associated with) dwelling-places of the dead. Probably other off-shore islands, too, have been conceived of as homes of the ancestors, although evidence of this has not survived down to us. Candidates for this would be islands that are conceived of in popular lore as dwellings of fairies or Otherworld powers in general – like Blå Jungfrun in Kalmarsund, South-eastern Sweden, and Seley off eastern Iceland.

Concluding remarks

I am not claiming that being beyond water is the only source for liminality nor that it is sufficient. *Differentness* is the quality that takes to make a holy place.[127] Being beyond water is not the only one way of being different and in a landscape with many islands it is not enough. Then, being different from other islands is required.

In this article I have gathered examples and indications that one may cross water – horizontally or vertically – to go to the Scandinavian gods, especially Óðinn. This has not been pointed out clearly before, especially not the vertical version. It probably has to do with the fact that there is no hint of such water crossing in Snorri's *Gylfaginning*[128] and the fact that it contradicts Snorri's version of Old Norse cosmology.[129] This is not a strong objection, though, because Snorri's account as well as the rest of the Old Norse corpus are full of contradictions about world view, as has been

---

126 Brink, "Mythologizing Landscape," 95.

127 Cf. Nathan Söderblom, "Holiness," in *Encyclopædia of Religion and Ethics* 6, ed. James Hastings (Edinburgh: T. & T. Clark, 1913), 731-41; C.W. von Sydow, "Det ovanligas betydelse i tro och sed," *Folkminnen och folktankar* 1926: 53-70; Veikko Anttonen, "Does the Sacred Make a Difference? Category Formation in Comparative Religion," in *Approaching Religion*. Part I. *Based on Papers Read at the Symposium on Methodology in the Study of Religions Held at Åbo, Finland, on the 4th-7th August 1997*, Scripta Instituti Donneriani Aboensis 17 (Åbo: Donner Institute for Research in Religious and Cultural History, 1999), 12.

128 *Edda Snorra Sturlusonar*, 8-77.

129 *Ibidem*, 15-6.

pointed out by Brink, Wellendorf and others.[130] Moreover, there is reason to believe that Snorri's presentation was influenced by Christian notions because he mostly places the gods in the sky. Still, some of our information clearly indicates that the idea of gods in the sky was a pre-Christian notion to some extent, too. However, it is thinkable that the contradiction is only seemingly between the accounts that seem to place the gods and the Otherworld underground, beyond or under water, in the sky, etc. In countless accounts, both from Scandinavia and other parts of the world, one comes to the same kind of Otherworld no matter what passage one passes through. If the passages are passages, not the Otherworld itself, the contradiction may be avoided. I intend to discuss this in a later article.

Acknowledgements: Thanks to Jonas Wellendorf, Andreas Nordberg, and the participants at the seminar at Utstein in June 2008 for comments on drafts of this paper.

---

130 Stefan Brink, 2004, "Mytologiska rum och eskatologiska föreställningar i det vikingatida Norden," in *Ordning mot kaos. Studier av nordisk förkristen kosmologi*, ed. Anders Andrén et al., Vägar till Midgård 4 (Lund: Nordic Academic Press, 2004), 291-316; Jonas Wellendorf, "Homogenity and Heterogenity in Old Norse Cosmology," in *Old Norse religion in long-term perspectives. Origins, changes, and interactions. An international conference in Lund, Sweden, June 3-7, 2004*, ed. Anders Andrén et al., Vägar till Midgård 8 (Lund: Nordic Academic Press, 2006), 50-3.

HVÍTRAMANNALAND AND OTHER FICTIONAL ISLANDS IN THE SEA

Else Mundal

Several Old Icelandic texts mention a place in the Atlantic Ocean called Hvítramannaland, "the land of white men." This land is mentioned in *Eiríks saga rauða*,[1] ch. 12; *Landnámabók*[2] (Sturlubók, ch. 122 and Hauksbók, ch. 94); and in *Eyrbyggja saga*,[3] ch. 64, a land – or island – in the same location, southwest of Ireland and close to Vínland, is mentioned.

Eiríks saga rauða tells how Karlsefni and his crew, on their way back from Vínland to Greenland, in a place which they called Markland – somewhere on the coast of North America – found five *skrælingar*, the Old Norse word for the natives the Nordic people met in Greenland and Vínland. Two of them were young boys, whom they caught and took back to Greenland, where they taught them the language and had them baptized. According to the saga, the boys told them that there was a land just opposite their land where there lived people who were dressed in white and, crying out loudly, carried poles in front of them to which pennants were attached. People thought, the saga says, that this must have been Hvítramannaland, or *Írland it mikla*, Great Ireland.

Landnámabók tells how an Icelander, Ari Másson, great-great grandfather of Ari inn fróði, drifted away and reached Hvítramannaland, which some people call *Írland it mikla*. This land is six days and nights of sailing west of Ireland, the source says. The distance, measured in how many days and nights of sailing it would normally take to get there, is a distance a little shorter than the distance between the west coast of Norway and the east coast of Iceland. That distance, according to the same source, *Landnámabók* (Sturlubók, ch. 2 and Hauksbók, ch. 2), is seven days and nights of sailing. The distance between Ireland and Hvítramannaland would then bring one to the middle of the ocean.

---

1 *Eiríks saga rauða*, ed. Einar Ól. Sveinsson and Matthías Þórðarson, in *Íslenzk fornrit* 4 (Reykjavík: Hið Íslenzka Fornritafélag, 1935).

2 *Landnámabók*, ed. Jakob Benediktsson, in *Íslenzk fornrit* 1 (Reykjavík: Hið Íslenzka Fornritafelag, 1968).

3 *Eyrbyggja saga*, ed. Einar Ól. Sveinsson and Matthías Þórðarson, in *Íslenzk fornrit* 4 (Reykjavík: Hið Íslenzka Fornritafélag 1935).

Hvítramannaland is, however, situated close to *Vínland it góða*, "Vinland the good,"[4] as is said in the story about Ari Másson. It is also said that Ari never succeeded in getting away from this island and he was baptized there. *Landnámabók* also mentions its sources for the story about Ari Másson, who ended his days in Hvítramannaland. The story was first told by a man called Hrafn Hlymreksfari, who had stayed for a long time in Limerick (Old Norse *Hlymreka*) in Ireland. Additionally, the source says that Þorkell Gellisson (the parternal uncle of Ari inn fróði) had said that Icelandic men had heard Þorfinnr in the Orkneys (earl of the Orkneys, who died in 1064) say that Ari Másson had been recognized in Hvítramannaland and he was held in high esteem there, but he did not succeed in getting away.

In *Eyrbyggja saga* we have the sad story about Bjǫrn breiðvíkingakappi and Þuríð, the sister of the chieftain Snorri góði. These two loved each other, but Snorri married his sister off to another man. They continued to meet, but after some time Bjǫrn had to flee the land. He returned after a few years, still very interested in Þuríðr and the young son she had given birth to in the meantime. He had to flee the land for a second time and afterwards no one heard about him for a long time, the saga says. After many years an Icelandic crew on a ship on their way back to Iceland drifted off course west of Ireland, in a southwesterly direction, far out to sea. After a long time they saw land – it was a large land, but they did not know what land it was. They sailed into a harbour and landed, and suddenly they were surrounded by a large crowd of men whom they thought might speak Irish. The men on the island were very unfriendly. They attacked them, captured them, tied them up, and brought them on land. They were taken to a meeting and they understood that some would kill them at once and others would make them slaves. Then an old man, followed by many others, came riding towards them, and the people at the assembly greeted him as their master. He spoke to the Icelanders in their own language and was very interested in news from Iceland and he was especially interested in Þuríðr and her son and sent gifts to them – a ring to Þuríðr and a sword to her son. He would, however, not tell his name, but people felt sure that this man was Bjǫrn. This man, whoever he was, advised the Icelanders to leave the place as soon as possible and he forbade anyone to try to find him because – as he said ·

> *...hér er land vítt ok illt til hafna, en ráðinn ófriðr alls staðar útlendum mǫnnum, ...*
>
> [...here large areas and harbours are few and far between and there is always strife for men from abroad everywhere ...]

It has been a matter of discussion whether the conception of an island somewhere far away to the southwest was rooted in someone's experience of having found land further

[4] Hauksbók has the form Vindland in this place, and Sturlubók writes it as Írland. The form Vínland is found in Skarðsbók, which is a later manuscript based on Sturlubók and Hauksbók.

to the south than Vínland.[5] The Norsemen had discovered many new lands – the Faroe Islands, Iceland, Greenland, and Vínland – so for them it must have been logical to think that there could be other lands far out in the sea.

It seems, however, that the conception of Hvítramannaland is more rooted in Irish tradition than in Old Norse experiences at sea. In both *Eiríks saga rauða* and *Landnámabók* the mysterious Hvítramannaland is called *Írland it mikla* with another name which clearly points in the direction of Ireland. In the story about Bjǫrn breiðvíkingakappi in *Eyrbyggja saga*, the Icelanders who found Bjǫrn thought that the people in this remote island, whom they could not understand, spoke Irish. In the story about Ari Másson in *Landnámabók*, the sources for the story are traced back to a man who had lived in Ireland for a long time and to Earl Þorfinnr in the Orkneys. In *Eiríks saga rauða* the two native boys who were brought back to Greenland are mentioned as sources for the existence of a land which people in Greenland identified with Hvítramannaland, thus making this land close to the lands discovered by the Greenlanders. The story about Ari Másson in *Landnámabók* mentions that Hvítramannaland is situated close to Vínland. This indicates that the conceptions of Hvítramannaland, which the Norsemen most likely picked up in Ireland before they found the east coast of North America, were connected to Vínland and thought to be situated somewhere in the same area after the discovery of the new lands in the west.

In Ireland a great tradition existed for stories about voyages at sea. One Irish genre, called *echtra*, described voyages searching for the Otherworld. Another genre, which was called by the Irish word *immram*, described voyages at sea. The *immram* was also used for stories – oral and written – about monks who explored the sea to find distant islands where they could settle down – in much the same way as holy men had settled in the desert in the time of the Primitive Church.[6] The Latin translation of the native genre term, *immram*, when applied to stories of seafaring monks, was usually *navigatio*. The "Christianized" *immram* – influenced by the *echtra* genre – became quite popular and the stories about St. Brendan (d. c. 575), who became especially famous for his sea voyages, were most renowned. The early development of the stories about St. Brendan is unknown, but from the tenth century onwards we have manuscripts of *Navigatio sancti Brendani abbatis*, a tale of the voyages of St. Brendan and his monks over the ocean

---

5 See William H. Babcock, *Legendary Islands of the Atlantic: a Study in Medieval Geography* (New York: American Geographical Society, 1922), 184.

6 On these genres, see Walter Haug, "Vom Imram zur Aventiure-Fahrt. Zur Frage nach Vorgeschichte der hochhöfischen Epenstruktur," *Wolfram-Studien* 1 (1970): 264-98; idem, "The Little Man on a Leaf and the Two Concepts of the *Dutch/German Reise*," in *The Brendan Legend. Texts and Versions*, ed, Glyn Burgess and Clara Strijbosch (Leiden: Brill, 2006), 81-98.

before reaching the paradise Land of the Saints.[7] The *Vita Brendani*, in which the saint's voyages at sea are also mentioned, exists in manuscripts from the twelfth century onwards.[8] It is, however, commonly believed that stories about St. Brendan's sea voyages were in circulation shortly after his death and that a legend about the saint, on which both the later *Vita* and the *Navigatio* were based, existed as early as the seventh century.

When the Vikings settled in Ireland and came into close contact with Irish culture they must have been acquainted with the *echtra* genre as well as with the secular and the "Christianized" *immram*. The Vikings, coming from a seafaring nation themselves, may have found these tales interesting even though they must also have seemed unfamiliar and strange in many ways. In the stories about St. Brendan – which exist and existed in different versions – the voyage brings the readers and the listeners to many strange places in the sea, for instance, to an uninhabited island with an empty town where everything is prepared for St. Brendan and his monks, to an island full of enormously large sheep, to an island which turned out to be a whale or a big fish, to an island inhabited by beautiful white birds that turned out to be the souls of the angels who had fallen together with Lucifer, to an island with a monastery and monks, to an island that was an enormously high pillar in the sea, to frightening islands close to Hell where devils could be seen, to a rock in the ocean to which Judas Iscariot was clinging, to an island where a hermit lived who had subsisted on fish that an otter brought him for 30 years and for the last 60 years had lived on healthy water alone, and finally, to the Paradise that once existed for Adam and Eve – where no living person was allowed to go ashore. In addition, we hear of fights between sea monsters, helping and talking animals, and other unbelievable things.

It seems reasonable that the descriptions of some of the islands may be based on real experiences at sea. It is, for instance, suggested that the island of sheep and the island of birds are two of the Faroe Islands, and that the description of islands close to Hell must have been based on an observation of volcanic eruptions in Iceland; there are in fact other sources as well which indicate that Irish monks had settled in those places before the arrival of the Norse settlers.[9] The Irish stories are, however, told in a non-realistic mode. In the Christianized *immram* the Christian miracles lend zest to the story, but there are also many elements in the Irish stories which must be characterized as fairy-tale motifs.

---

7 This text spread widely and was translated into most Western European vernaculars, also into Old Norse. The oldest extant Latin version was probably written outside Ireland, see Michaela Zelzer, "Philological Remarks on the So-called *Navigatio S. Brendani*," In *The Brendan Legend. Texts and Versions*, 337-350.

8 A survey of manuscripts and studies on St. Branden is found in Gly Burgess and Clara Strijbosch (ed.), *The Legend of St Brendan: A Critical Bibliography* (Dublin: Royal Irish Academy, 2000).

9 In *Íslendingabók* (ch. 1), written in the 1120s, Ari inn fróði mentions that the Norse settlers in Iceland found objects which Irish monks had left behind.

In Old Norse tradition, especially in texts of the *fornaldarsaga* type, the heroes can reach the land of trolls when crossing the ocean to the north and the northeast. Trolls can also be placed in Greenland – far to the north – but otherwise there are not many fantastic stories to be told about the seafaring Norsemen. They sail the seas, find land and cultivate it, and the hardships at sea are normally not dwelt upon. Since the saga genres – even the so-called realistic saga genres – are mixed modally, fantastic elements can be found in any saga, especially in stories which took place long ago and far away. In stories telling about sea journeys under distant skies some fantastic elements should be expected, even in the sagas of Icelanders, and they do in fact occur. In *Bjarnar saga hítdœlakappa*[10] (ch. 5), for example, Bjǫrn kills a flying dragon on a sea journey, and in *Eiríks saga rauða* (ch. 12) the Greenlanders find a creature who walks on one leg, a so-called *einfœtingr*. The sagas telling about the long sea voyages to Vínland are otherwise told in the same realistic mode as other sagas belonging to the same genre, and it is commonly believed that the places told about in *Grœnlendinga saga* and *Eiríks saga rauða* – Vínland, Markland and Helluland – are real places on the east cost of North America, even though they have not been located with certainty.[11]

What, then, about Hvítramannaland – is that land fact or fiction? The two sagas which tell about voyages to lands southwest of Greenland never mention that the Greenlanders came to a land which they named Hvítramannaland. The place is, however, mentioned in *Eiríks saga rauða* where the land the two young native boys told about is identified with Hvítramannaland. We must, of course, consider it a possibility that the conceptions of Hvítramannaland were at least partly based upon stories rooted in the Old Norse culture about seafaring men who drifted off and really reached land far away in a southwesterly direction, but as pointed out, the tradition of Hvítramannaland is so closely connected to Ireland both through the alternative name, *Írland it mikla*, Great Ireland, and through references to Irish tradition that it seems reasonable that the tradition of this mysterious place in the sea goes back to Irish traditions to an essential degree, but was adapted and transformed in an Old Norse context. The question is, what sorts of events or conditions in the Old Norse culture may have contributed to the Old Norse conceptions of Hvítramannaland?

Hvitramannaland is not – as the lands in the sea in the Irish *echtra* genre – a land belonging to the Other World. Hvítramannaland is inhabited by real mortal people. The only detail which perhaps points in another direction is that the story about Ari Másson could indicate that this land is a place of no return. The Irish *immram* genre, perhaps especially the Christianized form, may have left some marks. It is obvious that

---

10 *Bjarnar saga Hítdœlakappa*, ed. Sigurður Nordal and Guðni Jónsson, in *Íslenzk fornrit* 3 (Reykjavík: Hið Íslenzka Fornritafélag, 1938).

11 Owing to Inge Stine Ingstad, *Excavations at L'Anse aux Meadows, Newfoundland, 1961-1968* (Oslo: Universitetsforlaget, 1977) we know of a Norse settlement there.

Hvítramannaland is a Christian place. *Landnámabók* tells that Ári Másson, who lived before the Christianization of Iceland, was baptized there. In *Eiríks saga rauða* the two native boys described how in Hvítramannaland there lived people dressed in white who, crying out loudly, carried poles in front of them on which pennants were attached. In the edition of the saga in *Íslenzk fornrit*, the editor, Matthías Þórðarson, writes in a footnote to this story that the model for this description must be religious processions. The white colour could suggest Christian connotations, but the poles with pennants attached and the loud crying out could be interpreted in ways other than some sort of a religious procession. It is, in my opinion, equally likely that the poles with pennants attached and the loud crying out describe some sort of warlike behaviour, battle array and war cries. In that case the people of this land can be described as hostile towards intruders.

In the edition of *Landnámabók* in *Íslenzk fornrit*, the editor, Jakob Benediktsson, comments in a footnote on the information about Ari Másson that stories about fertile islands in the sea came to Iceland at an early stage.[12] That may well be true. The new world that rises from the sea in *Vǫluspá*, where fields grow without being sown, may build on conceptions of that kind, and the descriptions of Vínland in *Eiríks saga rauða* and *Grœnlandinga saga* may also have been influenced by conceptions of fertile and pleasant islands in the sea – even though the stories about wild grain may have a real background; it is commonly believed that what the Norsemen found was wild rice. The picture drawn in the different texts of Hvítramannaland is, however, not a picture of a fertile and pleasant island. On the contrary, this is a dangerous place. The place may be fertile, since Hvitramannaland could evidently produce enough food to support a large population, but this is not the focus of the texts. What is focused on is that a large, and therefore frightening, population is hostile towards strangers, whom they want to kill or keep as slaves. Only in exceptional cases, as in the cases of Bjǫrn breiðvíkingakappi and Ari Másson, would strangers be accepted and become members of the society.

What sort of experiences could have contributed to the conceptions of such an island in the sea? In my opinion the picture of an island (or a land) in the Atlantic Ocean with a hostile population who wanted to kill or enslave intruders reflects the Norsemen's meeting with the Native Americans, which may have been a more frightening experience than the sagas admit.

It is clear from both *Eiríks saga rauða* and *Grœnlendinga saga* that not all meetings with Native Americans were peaceful. The natives wanted to trade, but they also attacked and killed a few Greenlanders. The sagas admit that the Greenlanders had to take the threat from the natives seriously. *Eiríks saga rauða* (ch. 11) clearly tells that the Native Americans frightened Karlsefni and his people and he chose to go back to Greenland because he understood that the native people would cause them fear and

---

12 *Landnámabók*, 162, note 2.

strife. *Grœnlendinga saga* (ch. 7) describes how Karlsefni, after the natives had tried to enter his house, built a strong fence around the houses, and as soon as spring came he announced that he wanted to go back to Greenland.

The picture of Native Americans is, however, ambiguous. At the same time as the Old Norse sagas admit that the natives caused fear, they are depicted in a slightly comical light. They are afraid of a vicious bull which the Greenlanders had, and they are obviously fooled when trading with the Greenlanders. In *Eiríks saga rauða* the Native Americans exchanged fur for a strip of red cloth which they knotted around their heads – and the Greenlanders cut the strips narrower and narrower while at the same time raising the prices. In *Grœnlendinga saga* they exchanged fur for milk, butter, and cheese, and as the saga says, carried their goods away in their stomachs while the Greenlanders got plenty of good furs. In both sagas there is also a story which depicts the natives in a comical light when they reveal that they do not know how to use an iron axe. In *Eiríks saga rauða* (ch. 11) the natives found an axe and tried to cut a tree and were very satisfied, but then when they tried to cut a stone and the axe, not the stone, broke they lost interest in the axe. In *Grœnlendinga saga* (ch. 7) the axe was picked up by a young native who tested it by cutting one of his own men on the head, with, of course, a fatal result. The most comical scene is, however, found only in *Eiríks saga rauða* (ch. 11): When the Native Americans attacked, the Greenlanders had to flee and try to find a good place for defence. Freydís, the daughter of Eiríkr rauði, ran after the Greenlanders goading them, but since she was pregnant she was not able to run fast enough, and the Native Americans surrounded her and attacked. She had, however, picked up a sword which had belonged to one of the Greenlanders, and she drew forth one of her breasts from under her clothes and whetted the sword on her naked breast. That scared the natives greatly, and they ran away.

To depict the Native Americans humorously was probably necessary to prevent the Greenlanders from loosing their status as heroes. Reading between the lines in *Eiríks saga rauða* and *Grœnlendinga saga* it is obvious that the native people defended themselves and their land very well. These rather scary experiences in the new lands in the West are, in my opinion, the events and conditions which transformed the different conceptions about these – more or less – fictional islands in the sea in the older Irish and Nordic culture into the conceptions of Hvítramannaland: a dangerous place with a large and hostile native population who attacked and wanted to kill or enslave strangers; a land which no one should try to find, where it would be best not to go ashore, or if one had to go ashore, to leave as soon as possible.

THE UNKNOWN ISLAND

Margaret Elphinstone

In a short story by José Saramego called, "The Tale of the Unknown Island," a man goes to the king and boldly asks for a boat so that he can go in search of the unknown island:

> What unknown island, asked the king, suppressing his laughter as if he had before him one of those utter madmen obsessed with sea voyages, whom it would be as well not to cross, at least not straight away. The unknown island, the man said again, Nonsense, there are no more unknown islands, Who told you, sir, that there are no more unknown islands, They're on all the maps, And what is this unknown island that you want to go in search of? If I could tell you that, it wouldn't be unknown.[1]

The Unknown Island is a motif that constantly recurs in tales, written narratives, and maps pertaining to the Atlantic seaboard of Europe. Representations of the Unknown Island are necessarily ironic: How can an artist, in any medium, represent what is by definition beyond comprehension? And yet isn't this precisely what art sets out to do? An open-ended text is itself an act of exploration that takes us further than the limits of our conscious knowledge. The Unknown Island is, in fact, a self-reflexive motif. The sea voyage into uncharted territory, culminating in the discovery of the Unknown Island, mirrors the symbolic journey of writer and reader, propelled by the impetus of the story.

The provenance of European tales and texts focussing on the Unknown Island covers more than twenty-five centuries in time, and three thousand seven hundred kilometres (that's roughly from Lisbon to Narvik) in distance. I imagine the story of the Unknown Island may be almost as old as our human presence on Europe's west coasts. Certainly it appears to be equally widespread. José Saramego wrote his *Tale of the Unknown Island* in Portugal at the very end of the twentieth century; his provocative and appealing text is one of the latest re-tellings of a story so well-established in the Euro-

---

1 José Saramego, *O Conto da Ilha Desconhecida* (Lisbon: Assirio and Alvin, 1998); *The Tale of the Unknown Island*, tr. Margaret Jull Costa (London: Harvill, 1999) p 19.

pean west-coast psyche that it's perhaps no exaggeration to describe readerly response as literally atavistic.

So what has the Unknown Island signified through this long human history? What did the open-ended Atlantic mean to the peoples of the past who looked out onto its challenging waters? Why should more than two millennia of quest narratives be projected out into the open sea, even where no islands can be immediately discerned? What is the object of the sea-borne quest? What longings do people hope to have assuaged by its fulfilment?

On one level the answers are pragmatic. Real islands could offer potential land, resources, wealth, and sanctuary. When these practical needs were not met at home, there was always the possibility of more islands out there beyond the horizon. The story of Floki releasing his three ravens, and thus discovering Iceland, has the structure and ambience of folk tale – the portentous ravens, the magic number three, the quest for the Unknown Island – and, for a Christian/Judaic audience, the resonance with the doves that Noah released from the Ark. The story of Floki is constructed according to the conventions of myth.[2] But it's also a thoroughly practical account; experienced sailors did routinely use birds to indicate where land lay. The land-taking of Iceland was a very practical response to land hunger in Norway, and the need of a vigorous, war-like, demanding people to expand into new territory. Floki was a real person. He wasn't on an imaginary quest: he was seeking real land, real wealth, independence and opportunity. But again, the binary polarity breaks down even as it's expressed. Pragmatic opportunism and mythical quest are not so easily disentangled. Floki's journey lies in the debateable land where myth and fact continually wash over one another as perceptions rise and fall.

For as long as islands remain unknown, their ontological status is necessarily in doubt. Early debates about unknown islands focus largely on this debateable borderland between fact and myth. Like the space between tidelines around the shores of the islands of the Atlantic, the liminal space between mythical and real islands is shifting and uncertain. It doesn't wholly belong to one element or another, but lends itself to both. It's this space between fact and myth that I want to explore.

Modern approaches to unknown islands are inevitably illuminated by the eighteenth-century Enlightenment and its aftermath. Consider two apparently disparate manifestations of the Scottish Enlightenment; while scholars such as Hume and Reid were illuminating ancient philosophies and theologies with the relentless beam of rational analysis, Harrison was submitting his design for a ship's chronometer that would accurately measure longitude, and in Scotland the Stevenson family of engineers were literally enlightening the treacherous shoals and islands of the Scottish seas. Remote

---

2 Hermann Palsson and Paul Edwards (trans.), *Landnámabók* (Winnipeg: University of Manitoba Icelandic Studies, 1972), vol. 1, 17-18.

islands that had been havens of piracy, uncertainty, and mythic ambiguity, suddenly became accessible through the enlightening beams of logic, reason and trigonometry.

Anyone who's sailed at night among the islands of the western seaboard knows the reassuring power of lighthouses. You're tossing on dark waters, all you can hear is wind and sea, the stars (we're in the north Atlantic here) are obscured by cloud. You can't even stand upright – everything seems chaotic. You pick up a beam of light far away – just where it should be according to the compass and the chart. You take a bearing and look for the next one – there's a tiny pinprick of light in the right place. You breathe a sigh of relief. Thanks to the engineers of the Scottish Enlightenment, you know exactly where you are.

When Pytheas went looking for the mythical island of Britain, when Strabo described what lay beyond the Pillars of Hercules, when Columba and Brendan set sail in their skin boats, or the Norse Land-takers set out for the western horizon, there was no such orderly progression of lights. What did the sea feel like when the night brought nothing but darkness and chaos? What was the status of the unknown island then? Anyone, in any century, who goes to sea, often has to ask "is that a real island, or not?" One could call it the perennial searching question. We may know the island's there from the blip on the radar, but to this day we also do what medieval explorers did, although our observation skills are vastly inferior – we look for a stationary cloud hanging over an invisible strip of land or birds flying back to their nests or tell-tale shifts in tide or current. Even if modern technology tells us exactly what's out there and gives us every instrument available to navigate safely round it, I suggest that, as sailors, we have an atavistic desire to locate and define the unknown island using the evidence of our own senses. After all, our survival may depend on it. For pre-Enlightenment, pre-industrial seafarers, without radar, without longitude, without map, the ontological status of the unknown island was dangerously ambiguous.

One can read the trajectory of exploration of the islands of the western sea-board in two ways. One way is to see it as an expansion of knowledge and technology, as the physical geography of the northern and western regions was gradually opened up. The ability to measure longitude removed many imaginary islands from the map of the Atlantic in one stroke. (And yet the British Admiralty charts of 1854 still mapped islands that are not geographical entities. Symbolically, of course, Hy Brasil, like Atlantis, Tir Na Og or the Hesperides, is clearly located within our psyches to this day.) The second way is to see the trajectory north and west as a continual re-focussing of the unknown island myth.

Both views – scientific geography and mythic archetype – are open-ended. Physical exploration, correct mapping and geographical analysis reached the far side of the Atlantic long ago. Even so, empirical definitions have not ultimately located the Unknown Island. There are still no boundaries: chaos and the unknown continue to lurk beyond the beam of scientific enquiry. No one has plumbed the depths of the mid-Atlantic trench; all sorts of strange monsters are reputed to dwell in its unimaginable depths. No one can foretell exactly when the next volcano will erupt where the tectonic plates meet on the mid-Atlantic ridge. One day – tomorrow perhaps – the collapse of Gran Canaria will sweep away the east and west Atlantic seaboard in a new version of the Flood myth. Nothing the Enlightenment can produce is capable of subduing the fears and longings associated with the Unknown Island at the edges of our Atlantic-shaped consciousness.

In fact, the first view of exploration – locating and defining the Unknown Island as part of the expansion of scientific knowledge – is ultimately the same as the second view – the continual refocusing of the myth. We haven't shifted the ontological status of the Unknown Island after all. Maybe we haven't come as far as we thought.
The question: "Is that a real island or not?" has had many different meanings over time, besides the immediate practical definition of navigational hazard or chance of survival. Dualism, in the form of a perceived polarity between the material and spiritual, has hovered just beyond the boundaries of orthodoxy throughout the Christian era. In medieval Europe the spirit was the dominant partner in the binary opposition; the Church exercised social control by focussing beyond the temporality, injustices, and sensuality of physical life. It projected human attention onto an eternal, spiritual polarity based on the absolute separation of good and evil, only attainable after material death. Broadly speaking, in the modern world the weight has shifted towards the dominance of material realities over spiritual ones.

But what about the Unknown Island? It lies beyond the boundaries of geographical knowledge – or it wouldn't be – and yet it's real enough to be attainable in this world rather than in an equivocal heaven. You don't necessarily have to die to get to the Unknown Island; you just have to make a dangerous and uncharted sea voyage. You may not return, or if you do, it certainly won't be to exactly where you came from. In fact, reaching the island is going to be a sort of death after all, but not one located in any Christian hierarchy. Even the Christian Celtic monks knew that the white martyrdom – not death but self-imposed exile to the fringes of the world – was neither heaven nor hell; one could argue that it was redemptive precisely because it was both heaven and hell at once, encapsulated in the small space of the Unknown Island.

The Scottish ballad of True Thomas isn't about islands per se, but it locates a place set apart from orthodox hierarchies of good and evil. It isn't the sea that separates us from that dangerous yet desirable locus. It's magic: we embark from solid ground into the unpredictable, alien world of Faerie:

> Oh see not ye yon narrow road
> So thick beset with thorns and briars?
> That is the path of righteousness

Tho after it but few enquires.

Oh see not ye yon braid braid road
That winds across the lily leven?
That is the path of wickedness
Tho some call it the road to heaven.

And see not ye that bonnie road
That winds about the fernie brae?
That is the road to fair Elfland
Where thou and I this day maun gae.[3]

Like the road to Faerie in "True Thomas," the journey to the Unknown Island is a way of evading the spiritual hierarchies and orthodoxies of the mainland world. It's a way of avoiding the lethal polarisation of material and spiritual. Because the island is *always* unknown, it never ceases to belong to the world of myth. But because people have *always* known that there are real islands out there to the north and west, the unknown island has never ceased to be a geographical fact.

It's not hard to imagine how islands over the western horizon first gained their supernatural qualities. The blue stones of Rathlin Island, off the north coast of Ireland, or the prized bloodstone from Rum in the Hebrides, have been dispersed along the seaborne trade routes of the western seaboard since Mesolithic times. These islands, only intermittently visible from the mainland, provided unique commodities. Intrepid sailors must have voyaged off-shore in their skin boats or dugout canoes, vanishing into the mists, and coming back with stones to trade that couldn't be found anywhere more accessible. No doubt the sailing directions, and the landfalls, were closely guarded secrets within the clan. The stones themselves, being so rare, must have been charged with spiritual power. Islands like Rum and Rathlin were sources of very real treasure to our hunter-gatherer ancestors on the British Atlantic coasts. The shrouded origins of the stones must indeed have seemed magical.

Pytheas' Pretanicca (which we now call Britain) occupied a similar status, from the point of view of Mediterranean chroniclers, in the early Celtic Iron Age. Pytheas' purpose was to describe at first hand the sources of tin and amber, crucial imports to the Hellenistic colonies of the western Mediterranean. His written text "On the Ocean" is no longer extant, but later chroniclers who quote, refute, and deride his explorations nevertheless make his route moderately clear. Pytheas reached the tin mines of Cornwall. After circumnavigating and meticulously measuring the unknown island of Britain, he crossed over to the amber-strewn shores of the southern Baltic. He was also the first

---

3 Emily Lyle (ed.) *Scottish Ballads* (Edinburgh: Canongate 1994) p 133

Greek to reach, and to name, Ultima Thule. Whether he reached the northern coasts of Norway or crossed the open seas to uninhabited Iceland is still open to debate.[4]

Like all texts about Unknown Islands, "On the Ocean" appears to beg the most obvious question about all mythical islands: unknown to *whom*? This silence, the obvious gap in the text, surfaces in every text about inhabited mythical islands. Sometimes, of course, the Unknown Island is totally uninhabited. As far as humankind is concerned, the life of the uninhabited island is not only unknown but also unknowable, in the absence of any observer. Who can imagine an utterly uninhabited island? Iceland before the first arrival, perhaps an Irish monk blown off course – a Pacific atoll before the first Polynesian canoe appeared over the horizon – the silent Azores first visible through a Portuguese spy-glass? There is no one to observe, no one to record. An utterly uninhabited island can only be accessed momentarily by an act of the imagination. The mirage can never be followed up. It dies with the first footfall.

But Pytheas' unknown islands were not like that. The fabled source of tin was a prosaic and probably brutal mining community, probably owned by Celtic warlords and worked by slaves. Unless Pytheas had his own ship (Victorian antiquaries credit him with a whole fleet), someone skippered the ship which took him north up the Irish Sea, to the Isle of Man, across the Minch to Lewis, and round Cape Wrath to Orkney. The skipper of that ship has left us no word of his existence, but anyone who's sailed those seas must realise that he knew what he was doing. We'll never know who he was, but for him the islands west and north of Britain can hardly have been unknown. In other words, at the heart of the myth of the Unknown Island there is an ironic silence. This silence is pregnant with human knowledge and consciousness that is now lost to us. Perhaps it's the forgotten, inaccessible aspect of our own ancestry. Eighteenth-century Primitivism often tended to focus upon the frozen, uninhabited spaces of the far north, as a chilly counterpoint to the orthodox Edenic myth. Historical narrative is the prerogative of those who can write. They are the ones with authority to tell us which islands are, or are not, unknown.

Pytheas may have had narrative authority over the non-literate, Iron Age Celts of Pretannica, but he had strayed too far from the centre of his own civilization to maintain posthumous credibility there. Strabo, writing a generation later, castigates Pytheas for his over-vivid imagination. A land where air and water conjoin into a damp "miasma," or where the sea congeals like cold jelly, can only be a fable in poor taste to the sophisticated Strabo. For his Mediterranean contemporaries, Pytheas had travelled beyond the bounds of even imagined possibilities. Those of us who still inhabit Pytheas' unknown islands can read between Strabo's dismissive lines, and see that Pytheas was quite right. Not only does he describe northern mists and sea ice correctly, but also his

---

[4] Barry Cunliffe, *The Extraordinary Voyage of Pytheas the Greek* (London: Penguin, 2001); Christina Horst Roseman (ed.), *Pytheas of Massilia: On the Ocean* (Chicago: Ares Publishers, 1994).

measurements of latitude and the length of the British coasts are startlingly accurate given the means at his disposal.[5]

The author of the *Navigatio Sancti Brendani Abbatis*, on the other hand, can hardly be commended for mathematically reliable sailing directions. Brendan voyages into a sea of myth, among unknown islands that are embodiments of semi-Christian myths mingled with archetypal dreams. The three islands of Brendan's *Navigatio*, the Island of Sheep, the Island of Birds, and the Island of the Community of Ailbe, are clear descendants of the Celtic myth-islands to the west, sometimes there and sometimes not. When the brothers make landfall on the Island of Birds, they sail up a magical river that only just accommodates their boat. They reach a spring where:

> ...there was a tree of extraordinary girth and no less height covered with white birds. They covered it so much that one could scarcely see its leaves or branches. When the man of God saw this, he began to think and ponder within himself what it means...[6]

The magic tree at the heart of the islands has its roots firmly set in Celtic and Norse myth. Whether it's an apple tree from the Hesperides or Tir Nan Og, or the ash tree Yggdrasil, is irrelevant. We recognise the branches weighed down with wonderfully animate life. Brendan's Island of Birds takes us close to the heart of Creation. Brendan's ensuing conversation with the bird could have come straight from the *Mabinogion*. The resonances with the magic cauldron that allows the hero to speak with the animals, with bird-spirits, and the female deities of tree and well, scarcely need elaborating here. The point is that the voyage of St. Brendan takes us to the very heart of the Unknown Island. Every time we reach the epicentre of the magical island space, the narrative neatly sidesteps into Christian doctrine. But no amount of sermon can hide the compellingly subversive nature of Brendan's islands.

The (literal) charm of Brendan's voyage is that it takes us into liminal spaces. It belongs to the same no-man's land, neither transcendent heaven nor material earth, as the land of Faerie in True Thomas. The twenty-first century reader of Brendan's *Navigatio* may well employ identical strategies to those used to enter the mythic world of another great mythic/Christian epic: Bunyan's *The Pilgrim's Progress*. The overt Christian

---

5 Cunliffe (*Pytheas*, 96-98), converts Pytheas' measurement of the length of Britain as 20,000 stades into kilometres. Depending on the stade length adopted, Cunliffe calculates Pytheas' figures at 7,400 km or 7,100 km. The modern *Encylopaedia Britannica* gives the length of coastline as 7,580 km. Cunliffe also offers detailed analysis and figures for Pytheas' measurements of latitude (see fig., p 124). He suggests that Pytheas probably took sun sights at Tregor, the Isle of Man, Lewis, Shetland, and, most controversially, at Thule. Cunliffe concludes (p. 98) 'Even for a highly experienced sailor the results are almost unbelievably accurate. The answer may simply be that all the inherent errors in his calculations tended to cancel each other out.'

6 John J O'Meara (ed. and tr.), *The Voyage of Saint Brendan: Journey to the Promised Land* (Gerrards Cross: Colin Smythe, 1978), 20.

sermonising is like a submerged reef, barring access to the mythic islands that delight the pagan soul.

It is curious that the power of the St. Brendan myth has been demonstrated within living memory by an adventure which replicates the original voyage, driven by the desire to make it *real*. Tim Severin's text, *The Brendan Voyage*,[7] describes an actual voyage made in a skin boat, derived from the Irish curragh, which the author argues is a replica of the boat in which Brendan and his monks must have put to sea. Severin and his crew follow what Severin argues is Brendan's original route, with the intention of proving that the *Navigatio* is a kind of hyperbolic log book of an Atlantic crossing from the west of Ireland, via Lewis, Faroe, and Iceland to Newfoundland.

Severin's voyage was itself the precursor of many more voyages made by intrepid young men, mostly from the northern coasts that were once prototypes of unknown islands, in boats that replicate the highly seaworthy but desperately uncomfortable open boats in which their ancestors – prehistoric, Celtic or Norse – once set sail because they had no choice.

Perhaps we can best return to José Saramego's twentieth-century short story of the Unknown Island via these real twentieth-century explorers in their retrospective skin boats. Severin's voyage is presented to us as a calculated experiment, undertaken as an empirical enquiry in order to prove a hypothesis. In other words, it pretends to be the product of the Enlightenment. I would argue that it is nothing of the sort. Undoubtedly the voyage is successful: *Brendan* finally makes landfall in Newfoundland with all the crew alive and well. The book is also a success: it re-tells vividly what must be the most ancient story told by dwellers on the Atlantic seaboard: the myth of men who venture into the open sea, and not only survive, but achieve all their hopes. The book ends appropriately with the celebration of this endlessly repeated, mythic adventure. As with the original Brendan and his monks, a spiritual journey has been completed. The characters have undergone a sea change, and the readers' perceptions too have been refined and altered. The resolution of the hypothesis that the eighth-century *Navigatio* actually describes an identical voyage to that undertaken by Severin and his crew is irrelevant. The fact is that we still have no proof whatsoever that Brendan and his monks followed this route or made landfall in North America. There has been no measurable scientific experiment. What *has* been replicated, one of the latest in a succession of more tellings than we can possibly imagine, is the myth about the search for the Unknown Island.

The antiquity of the Unknown Island myth is unprovable – I'd hazard my guess that it's as ancient as the communities which established themselves along our coast in the wake of the retreating ice[8]. Anyone who lives in view of offshore islands

---

7 Timothy Severin, *The Brendan Voyage* (London: Hutchinson & Co., 1978).

8 I have hinted at this in my novel *The Gathering Night* (Edinburgh: Canongate 2009) set on the west coast of Scotland in c. 6100 B.C.

knows how they can vanish for weeks on end, and then appear so close that they seem only a stone's throw away. On calm frosty days in winter sounds carry across the water as if they were coming from the next field. On some days the sea looks calm enough to swim across, and it's easy to row or paddle; on other days there is a terrifying barrier of white waves breaking right across the sound. "Near" and "far" cease to have any rigid definition. "There" or "Not there" is no longer a simple binary opposition. Even in a material world whose rules are bounded by numbered measurements and empirical reasoning, the senses are constantly deceived. We can't eschew the myth, even if we'd like to: the island insists on being unpredictable. However much we know about it, however often we've actually been there, as one looks at it across the water it remains elusive. It's constantly different. It's always keeping something in reserve, unknown.

In José Saramego's story the Unknown Island is discovered without the boat – or the reader – ever leaving harbour. In effect, the hero's voyage is about perception. Fulfilment is right next to him all the time, and all he needs to do is discover that it's there. The Unknown Island isn't as far away as he thought it was, but discovering its presence doesn't mean that he knows all about it. He finds it, and that makes him happy. But the text remains full of gaps. There are secrets. It would be a very unsatisfactory story if it were not so. However much we may want to have everything explained, it's even more important that the island, even in our demanding era, should remain unknown.

WILL THE SON OF NINE SISTERS RULE THE SEA-KIDNEY?

Bernt Øyvind Thorvaldsen

According to *Laxdæla Saga*, a wedding feast took place in Hjarðarholt on Iceland a couple of decades before the Icelanders officially converted to Christianity. Óláfr Hǫskuldsson, the bride's father, had built a splendid new hall with wall carvings:

> One of the guests at the feast was Ulf Uggason, who composed a poem about Olaf Hǫskuldsson and the carved legends depicted in the hall, and this poem he recited at the feast; it is called the 'House Lay', and is an excellent poem.[1]

The few stanzas of *The House Lay* which are preserved are not quoted in the saga, but in the thirteenth-century *Prose Edda*, a book on pagan mythology and vernacular poetry originally authored by Snorri Sturluson, but later reworked in manuscript transmission. The parts of *The House Lay* which are quoted in the *The Prose Edda* refer to well-known stories about the god Þórr, who fishes the world serpent, and about the funeral of Baldr, son of Óðinn.

One stanza, however, presents a more obscure myth about a fight between the god Heimdallr and the mythological trickster figure, Loki, over something which the poet calls a "sea-kidney" (*hafnýra*). In a section of *The Prose Edda* which presents different aspects of the god Heimdallr, it is said that he travelled to Vágasker and Singasteinn and struggled with Loki over "the ring of the Brisings" (*Brísingamen*), a precious necklace belonging to the goddess Freyja. *The Prose Edda* continues: "The skald Ulf Uggason devotes a lengthy passage to that story in his poem *Husdrapa*, and it is stated there that Heimdall and Loki took on the shape of seals."[2] The stanza of *The House Lay* which

1 Magnus Magnusson and Hermann Pálsson (trans.), *Laxdæla Saga* (London: Penguin Books, 1969), 112; "Þar var at boði Úlfr Uggason ok hafði ort kvæði um Óláf Hǫskuldsson ok um sǫgur þær, er skrifaðar váru á eldhúsinu, ok fœrði hann þar at boðinu. Þetta kvæði er kallat Húsdrápa ok er vel ort" [Einar Ól. Sveinsson (ed.), *Laxdæla saga*, Íslenzk fornrit, vol. 5 (Reykjavík: Hið íslenzka fornritafélag, 1934), 80].

2 Jesse L. Byock (trans.), *The Prose Edda.* (London: Penguin Books, 2005), 110; "Úlfr Uggason kvað í Húsdrápu langa stund eptir þeiri frásǫgu; er þess þar getit er þeir váru í sela líkjum" [Anthony Faulkes (ed.), *Skáldskaparmál*, vol. 1, Snorri Sturluson: Edda (London: Viking Society for Northern Research, University College London, 1998), 19].

deals with this myth is quoted later in *The Prose Edda*, and below is the first half of the stanza:[3]

Ráðgegninn bregðr ragna
rein at Singasteini
frægr við firna slœgjan
Fárbauta mǫg vári …

Prose order

Ráðgegninn frægr ragna reinvári bregðr við firna slœgjan mǫg Fárbauta at Singasteini…

Any reading of the stanza is based on philological choices, but for the sake of simplicity I will present my own interpretation of this stanza and the readings which are most relevant to it, without giving a full account of the scholarly history or competing interpretations.[4] In the first half-stanza, quoted above, there is a character called the "wise and famous border-guardian of the gods" (*ráðgegninn frægr ragna reinvári*), who is certainly Heimdallr, since he is considered the guardian of the gods in other mythological sources. The character called "the terribly sly son of Fárbauti" (*firna slœgr mǫgr Fárbauta*) is obviously Loki, whose slyness is well-known and who is the son of the giant Fárbauti. Although the identity of both characters is clear, what they actually *do* is more debatable. Kurt Schier gives an overview of previous interpretations of the expression *bregða við*, but suggests that it should be read as "turn against" or "fight."[5] Thus, Heimdallr, according to Schier, turns against or fights Loki at a place called *Singasteinn*.[6] In the second half-stanza, Loki is not mentioned at all:[7]

móðǫflugr ræðr mæðra
mǫgr hafnýra fǫgru,
kynnik, áðr ok einnar
átta, mærðar þǫttum.

---

3 Finnur Jónsson (ed.), *Den norsk-islandske skjaldedigtning* (Copenhagen: Gyldendal, 1912-1915), vol. I, B, 128.

4 For other interpretations, see Edith Marold, "Die *Húsdrápa* als kosmologisches Gedicht." In *Old Norse myths, literature and society: proceedings of the 11th International Saga Conference*, ed. Geraldine Barnes and Margaret Clunies Ross (Sydney: University of Sydney, 2000), 290-302. See also Sebastian Cöllen, "Der Ursprung des Feuers in nord-germanischer Mythologie. Studien zu *Húsdrápa* 2." *Arkiv för nordisk filologi* 122 (2007): 59-78.

5 Kurt Schier, "Húsdrápa 2: Heimdall, Loki und die Meerniere." In *Festgabe für Otto Höfler zum 75. Geburtstag*, red. Helmut. Birkhan (Vienna and Stuttgart: Wilhelm Braumüller, 1976), 577-88, at 579-80.

6 There are numerous and differing interpretations of *bregða við*. Marold suggests that Heimdallr "takes land away from" Loki (*bregðr rein við*), and relates this to a wider comparative material ("Die *Húsdrápa* als kosmologisches Gedicht", 295-8). I stick to the association of *rein* with *vári* in my reading, since the noun *rein* – in its meaning 'borderland" – suits very well the conception of Heimdallr as a watchman located on the border against giant territory.

7 The stanzaic representation is from Finnur Jónsson, *Den norsk-islandske skjaldedigtning*, vol. 1, B, 128, while the prose order is mine.

Prose order

... *áðr móðǫflugr mǫgr átta ok einnar mœðra ræðr hafnýra fǫgru; kynnik mærðar þǫttum.*

Heimdallr is referred to as "the powerful son of eight and one mothers" (*móðǫflugr mǫgr átta ok einnar mœðra*). According to several sources, the guardian god, Heimdallr, was brought into the world by nine women, a myth which will be treated in more detail below. The poet Úlfr says that Heimdallr rules (*ræðr*) the beautiful sea-kidney (*hafnýra*), and closes the stanza by proclaiming that he makes this known in a part of his poem (*kynnik mærðar þǫttum*).

What, then, does the "sea-kidney" refer to? Some scholars claim that the "sea-kidney" refers to the precious necklace which is mentioned in *The Prose Edda*, referred to above. The problem with this interpretation is that the metaphor (or *kenning*) itself does not easily support such an interpretation. On the contrary, as Schier points out, the "sea-kidney" is an expression which belongs to a group of *kennings* denoting rocks, cliffs or islands, for example "the heart of the sea" (*lagar hjarta*), "fjord-apple" (*fjarðepli*) and "the grain of the well" (*brunns bygg*). In these cases the shape of a heart, an apple or a grain is associated with water and sea, to denote cliffs, rocks or islands. In the same way, "sea-kidney" combines the form of a kidney with the ocean, and certainly belongs to this specific group of metaphors. Furthermore, Schier suggests that the sea-kidney is an island, presumably the same island which is called Singasteinn in the first half-stanza – the island on which Loki and Heimdallr are fighting. Since Heimdallr rules this island in the second half-stanza, the object of the struggle seems to be the island itself.[8]

In Schier's opinion, Singasteinn is the world as it arose from the sea in the times of creation. He suggests that we deal with an otherwise unattested creation myth in which Heimdallr and Loki fight to gain control over the world at the beginning of time. Comparable myths are widely attested throughout the world and represent an archaic mythological pattern.[9] Edith Marold also develops the possible comparative background to the motif.[10] In Old Norse mythology, there is, however, evidence of two mythological situations in which the world sinks and rises from the sea. The Eddic poem *The Seeress's Prophecy* (*Vǫluspá*) states that Óðinn and his brothers lifted the lands during creation (stanza 4):[11]

Before Burr's son	*Áðr Burs synir*
lifted up seashores,	*biǫdum um ypðo,*

8 Schier, "Húsdrápa 2", 583.

9 Schier, "Húsdrápa 2", 586; Kurt Schier, "Loki og Heimdallur: Athugasemdir um eðli og uppruna tveggja torræðra goða." In *Heiðin minni: Greinar um fornar bókmenntir*, ed. Haraldur Bessason and Baldur Hafstað (Reykjavík: Heimskringla, 1999), 25-46.

10 Marold, "Die Húsdrápa als kosmologisches Gedicht," 297-300.

11 Both the Old Norse text and the English translations of *The Seeress's Prophecy* are quoted from Ursula Dronke, ed. and trans., *Mythological Poems*. The Poetic Edda, vol. II (Oxford: Clarendon Press, 1997), 7-24.

they who moulded	*þeir er miðgarð*
glorious Miðgarðr.	*mæran skópo.*

Then, when the poem has presented the period in which Óðinn is in power, and the final destructive battle between gods and giants, the poem says that the world sinks into the ocean (stanza 54):

The sun starts to blacken,	*Sól tér sortna,*
land sinks into sea,	*sígr fold í mar,*
the radiant stars	*hverfa af himni*
recoil from the sky.	*heiðar stjǫrnor.*

Thereafter, the green world rises from the ocean for the second time (stanza 56):

She [the seeress] sees come up	*Sér hón upp koma*
a second time	*ǫðro sinni*
earth out of ocean	*iǫrð ór ægi*
once again green.	*iðiagræna.*

Thus, in *The Seeress's Prophecy,* the world rises from the ocean twice, first in the times of past creation and secondly in the future, after the world destruction. Heimdallr and Loki, however, are not mentioned in any of these cases, and the lifting of the lands in the times of creation is specifically associated with the sons of Borr, namely, Óðinn and his brothers.

The existence of a myth about Óðinn and his brothers, who lifted the world in the remote and creative past, does not exclude the possibility that the poet Úlfr presented a conflicting myth about Loki and Heimdallr in *The House Lay*. The source material of Old Norse mythology is heterogeneous and one should expect to find different and conflicting myths. But *The Seeress's Prophecy* opens another possibility: If *The House Lay* describes a fight over the world as it rises from the sea, then it could perhaps refer to a myth of future creation, when the world will rise from the sea again? In fact, there are few reasons to presume that the myth told in the stanza from *The House Lay* is placed in the remote cosmogonic past. The only word which might indicate this is *áðr*, which is either an adverb or a conjunction.[12] The adverb points to something which happened "earlier" or "a long time ago," and Schier naturally chooses to understand *áðr* in this way.[13] Thus he has some support for placing the myth in the past. I will suggest that *áðr* is the conjunction meaning "before" or "until," and that it introduces a subordinate clause attached to the sentence in the first half-stanza. Thus, I suggest the following reading of the whole stanza: "The wise and famous border-guardian of the gods

12 See *Ordbog over det norrøne prosasprog. A Dictionary of Old Norse Prose*, vol. 1 (Copenhagen: The Arnamagnæan Commission, 1995), 45-9.

13 Schier, "Húsdrápa 2", 580-1.

fights with the terribly sly son of Fárbauti at Singasteinn, until the powerful son of eight and one mothers rules the beautiful sea-kidney; that I proclaim in this section of the poem."[14] So, Heimdallr and Loki fight at Singasteinn until Heimdallr rules the island. Schier's substantial comparative material suggests that this island might be the world, but on the background of *The Seeress Prophecy* and my reading of *áðr* it is relevant to investigate if it makes sense to suggest that the myth could refer to the future, to the world which rises *after* the world is destroyed.

Mythology

When studying Old Norse myths one has to be aware of the heterogeneity of the source material. As I mentioned above, we expect conflicting myths, of different temporal, geographical and social origins. This situation makes it difficult to presume one mythological "system" behind all Old Norse myths. We must remember that "logic" in mythology and connections between different myths only exist in the minds of different individuals. At the same time, myths are social and traditional narratives and the conceptions of gods and supernatural forces will to some extent be shared within one social context. Although myths change over time and adapt to developments in society, it is also reasonable to presume *some* continuity in myths and mythology within a society.

With this in mind, one can approach the meaning of the second stanza in *The House Lay* by taking the mythological connotations in the stanza as a starting point. The poet Úlfr Uggason refers to the antagonists, Loki and Heimdallr, not by their names but by other characteristics. In the first half-stanza Heimdallr is considered wise in council (*ráðgegninn*), famous (*frægr*), and the guardian of the gods' borderlands (*ragna reinvári*); the second half-stanza describes him as powerful in his mind or by his bravery (*móðǫflugr*) and also states that he is the son of nine mothers (*mǫgr átta ok einnar mæðra*). Loki is described as terribly sly (*firna slægr*) and as the son of Fárbauti (*mǫgr Fárbauta*). It is clear that Úlfr's sympathy is in favor of Heimdallr, while his antipathy is reserved for Loki. We can also observe that Úlfr must have known at least three specific myths: the myth of Heimdallr's birth by nine women, of Fárbauti's fathering of Loki, and of Heimdallr's role as a guardian god. The other characteristics are more general and difficult to associate with specific myths.

These poetic connotations are important for two reasons: First of all, these myths and features are safe to include in the analysis of the stanza, since the poet both

---

14 Although it is common that half-stanzas in *dróttkvætt* poetry contain independent clauses, there are numerous examples of subordinate clauses introducing the second half-stanza. An example with *áðr* is a stanza attributed to Egill Skallagrímsson (Finnur Jónsson, *Den norsk-islandske skjaldedigtning*, vol. 1, B, 47, no. 21).

knew them and combined them himself. Secondly, if the poetic expressions referring to Heimdallr and Loki are not mere coincidence, but the result of conscious poetic creativity, it seems relevant to suggest that they might be keys to understanding the narrative of the stanza (the myth about how Heimdallr fights Loki to gain control over the island Singasteinn). If that is the case, we should now take a closer look at the subject of ancestry and kinship, since Úlfr emphasizes the ancestry of the two mythological antagonists in his poetic references.

Kinship

Although Úlfr himself associates the myth about the sea-kidney with the ancestry of the combatants, we still need to relate the information given about Heimdallr's and Loki's parentage to a wider context. First of all, norms associated with family relations are social phenomena, as all norms are. Kinship norms formed an important mental framework in medieval Scandinavia, both before and after the conversion to Christianity. Secondly, kinship norms play an important role in many Old Norse myths, as they do in Old Norse literature generally. Hence, I assume that Úlfr Uggason and his audience related the kinship of Loki and Heimdallr to norms existing in the social context. To some extent we need to presume that myths in other sources are somehow related to comparable conceptions of kinship as those present in the minds of Úlfr and his audience at Hjarðarholt in the 980s (in 983, according to Finnur Jónsson[15]).

In the Old Norse tales of the creation, Óðinn and his brothers are the dominant active forces. They kill a monstrous giant named Ymir, the progenitor of the giants. The gods then build the world from his body parts, according to *Grímnir's Sayings* (*Grímnismál*), stanza 40-41:

From Ymir's flesh the earth was made, *Ór Ymis holdi/ var iǫrð um skǫpuð,*
and from his blood, the sea, *en ór sveita sær,*
mountains from his bones, trees from his hair, *biǫrg ór beinom,/ baðmr ór hári*
and from his skull, the sky. *en ór hausi himinn.*

And from his eyelashes the cheerful gods *En ór hans brám/ gerðo blíð regin*
made earth in the middle for men; *miðgarð manna sonom;*
and from his brain were the hard-tempered clouds *en ór hans heila/ vóro þau in*

15 Finnur Jónsson, *Den norsk-islandske skjaldedigtning*, vol. 1, B, 128.

harðmóðgo
all made.[16] *ský ǫll um skǫpuð.*[17]

Thus the gods not only construct the world – they sow the seeds of their own destruction, since the murder of Ymir defines all his giant descendants as deadly enemies, and giants are powerful beings that in the end overturn the reign of Óðinn.[18] This hostile development is a mythological feud, comparable to feuds between human heroes in the saga literature that are heavily dependent on kinship relations. This point is underlined in two important books on Old Norse mythology: John Lindows' *Murder and Vengeance among the Gods*[19] and Margaret Clunies Ross' *Prolonged Echoes*.[20]

According to theories on Old Norse kinship, the loyalty of an individual was mainly defined by his patrilineal descent. Preben Meulengracht Sørensens describes two types of heroes in Old Norse literature based on two different types of exogamic relationships.[21] *The positive hero* is the result of an acceptable exogamic relationship: The mother is on a lower social level or an outsider while the father is of higher birth and integrated in society. The positive hero becomes an unusually powerful being and he his loyal to the kin of his father and thus serves society. Óðinn and his brothers are probably positive heroes, since they are the sons of a giantess (Bestla), and since their father (Borr) should probably be considered a god.[22] This origin explains Óðinn's power, the power which enables him and his brothers to create the world. At the same time, it means that Óðinn and the other gods fight their own matrilineal kinsmen, the giants, so that the whole mythological feud is between kinsmen. Although patrilineal kinship is the most important, the murder of matrilineal kinsmen – like Ymir and the other giants – is a negative and destructive action in an Old Norse context. Thus, the creation has a tragic aspect; the world is based on the shaky foundations of the murder of a kinsman.

While Óðinn and his brothers are positive heroes, the result of an acceptable exogamic relationship, the opposite relationship also exists. If a male outsider and a

---

16 Carolyne Larrington, trans., *The Poetic Edda*. Oxford World's Classics (Oxford: Oxford University Press, 1999), 57.

17 Jón Helgason, ed., *Eddadigte II: Gudedigte*. Nordisk filologi, vol. A, 7 (København, Oslo, Stockholm: Ejnar Munksgaard, Dreyers forlag, Svenska bokförlaget, 1961), 19-20.

18 See, however, Else Mundal's article "Androgyny as an Image of Chaos in Old Norse Mythology" (*Maal og Minne*, 1998, no. 1: 1-9) in which she argues that Ymir was considered a detestable and "chaotic" being because of his androgyny.

19 John Lindow, *Murder and Vengeance among the Gods: Baldr in Scandinavian Mythology*. FF Communications, vol. 262 (Helsinki: Suomalainen tiedeakatemia, 1997).

20 Margaret Clunies Ross, *The Myths*. Prolonged Echoes, vol. 1 (Odense: Odense University Press, 1994).

21 Preben Meulengracht Sørensen, "Starkaðr, Loki og Egill Skallagrímsson," in *Sjötíu ritgerðir helgaðar Jakobi Benediktssyni 20. júlí 1977*, ed. Jónas Kristjánsson and Einar G. Pétursson (Reykjavík: Stofnun Árna Magnússonar á Íslandi, 1977), vol. 2, 759-68.

22 Byock, *The Prose Edda*, 15; Anthony Faulkes, ed., *Gylfaginning*. Snorri Sturluson: Edda (London: Viking Society for Northern Research, University College London, 1988), 11.

woman of higher origin procreate, the result will be a *negative hero*. This is an unacceptable exogamic relationship, and the loyalty of the hero is, as it tends to be, bound to the father. Since the father is an outsider, the negative hero will turn against society and become a powerful, but destructive, force. Meulengracht Sørensen argues that Loki, the one who battles with Heimdallr over the "sea-kidney," is a negative hero – he was born by a goddess (Laufey) and his father is the giant Fárbauti. Thus, it is not surprising that Loki, who in many myths is included among the gods, turns out badly in the end. When the destruction of the world approaches and the last great battle between giants and gods occurs, he appears on the side of the giants.

Almost all of the gods whose ancestry is mentioned in the Old Norse sources are associated with either gods or giants through patrilineal kinship. Hence, their fate is bound to the destructive conflict which finally destroys the world. They *have to* participate in the final battle, and most of them die before the world finally sinks into the ocean. All beings with a giant or divine father must participate in the conflict due to the inescapable and momentous obligations towards their patrilineal kin.

What then about Heimdallr – whom the poet refers to as the "son of nine mothers" – how can his origin be interpreted according to Old Norse norms of kinship? In *The Prose Edda* Snorri Sturluson quotes a poem he calls *Heimdallr's Chant* (*Heimdallargaldr*). In the only two lines preserved, Heimdallr himself is speaking:

Of nine mothers I am the child,	*Níu em ek mœðra mǫgr,*
of nine sisters I am the son.[23]	*níu em ek systra sonr.*[24]

There are other sources referring to this myth, but the most extensive account is present in *The Song of Hyndla* (*Hyndluljóð*), in the 14th-century folio manuscript *Flateyjarbók*.[25] In ancient times, stanza 35 narrates, a boy was born to nine giant mothers at the edge of the world:

One was born in bygone days,	*Varð einn borinn / í árdaga,*
with enormous power of the sons of men;	*rammaukinn miǫk, / rǫgna kindar;*
then nine women gave birth to him, to the	*níu báru þann, / nad[d]gǫfgan mann,*
spear-magnificent man,	*iǫtna meyiar / við iarðar þrǫm.*[26]
daughters of giants, at the edge of the earth.[27]	

[23] Byock, *The Prose Edda*, 37.

[24] Faulkes, *Gylfaginning*, 26.

[25] GKS 1005 fol.

[26] Jón Helgason, *Eddadigte II: Gudedigte*, 86.

[27] Larrington, *The Poetic Edda*, 258.

The poem continues to enumerate the names of these nine mothers and then informs us that Heimdallr was strengthened with the power of earth, the ice cold sea, and boar's blood.

Some manuscripts of *The Prose Edda* claim that Óðinn was the father of Heimdallr.[28] Apart from the fact that no other sources confirm the existence of a father, there are other reasons to doubt this information. First of all, the poetic sources which refer to Heimdallr in kinship terms refer to him by his nine mothers. Why did the poets without exception refer to the matrilineal descent of the god instead of the more important patrilineal descent? In a patrilineal society it is natural to present a character by his *male* parentage. Thus, it is probable that poets like Úlfr Uggason did not associate Heimdallr with a father. The fact that Heimdallr has *nine* mothers points in the same direction, since nine was often considered a symbol of totality and completeness in the Old Norse context, and symbolically excludes a father. Furthermore, the idea of Heimdallr being a son of Óðinn fits too well into 13th-century euhemeristic reinterpretations of Old Norse mythology, where Óðinn is the human father of the other (human) gods. Thus, I doubt that the information about Heimdallr as a son of Óðinn should be given much weight in the reading of *The House Lay* from the 980s.

What, then, does the myth about Heimdallr's birth mean? According to theories on Old Norse kinship, the loyalty of an individual is defined patrilineally, as already stated. The absence of a father could mean that Heimdallr was not bound by the fateful laws of patrilineal loyalty. Unlike other mythological beings with a giant or god as a father, he is not tied to the unavoidable loyalty to patrilineal kin. In fact, the myth of Heimdallr's birth might mean that his role is peripheral in the imperfect and self-destructive world which Óðinn and his brothers established by the murder of Ymir. Let us now turn to the major role which Heimdallr is associated with in the world of Óðinn: his role as guardian or watchman. What can this function tell us about the god's position within Óðinn's world order?

The guardian god

The Eddic poem *Grímnir's Sayings* enumerates the residences of the gods, and in the thirteenth stanza, mentions Heimdallr's residence *Himinbjǫrg*:

Himinbiorg is the eighth, and there, they say,	*Himinbiǫrg ero en átto/ en þar Heimdall*
Heimdall rules over his sanctuaries;	*kveða valda véom;*
there the glad watchman of the gods drinks good mead	*þar vǫrðr goða/ drekkr í væro ranni*

[28] Finnur Jónsson, ed., *Edda Snorra Sturlusonar* (København: Gyldendal, 1931), 99.

in the comfortable hall.[29]	*glaðr [inn] góða miǫð.*[30]

What does the guardian role involve? *The Prose Edda* describes the role as follows: "He [Heimdallr] lives near Bifrost at a place called Himinbjorg. He is the watchman of the gods and sits at heaven's end, where he keeps watch over the bridge against the mountain giants."[31] *The Prose Edda* also informs us that Heimdallr possessed a horn called *Gjallarhorn*, "and its blast can be heard in all worlds."[32] When *The Prose Edda* reaches to the description of the world destruction, it narrates that monsters and giants will approach the lands of the gods. "As these events occur, Heimdall stands up and blows the Gjallarhorn with all his strength. He wakens all the gods, who then hold an assembly."[33] A comparable scenario is present in the Eddic poem *The Seeress's Prophecy*, which in this case might be the dominant source for Snorri's presentation.

Although the god is described as "powerful and sacred" in *The Prose Edda*,[34] his role as guardian is presented in a somewhat different perspective in an Eddic poem called *Loki's Quarrel* (*Lokasenna*). The gods are assembled in the hall of Ægir, a sea giant, and a quarrel is initiated by Loki. Insults are thrown between Loki and the gods, grave insults of perversity and unmanly and unwomanly behaviour. When Heimdallr addresses Loki and claims that he has lost his mind from drinking, Loki responds with the following insult, in stanza 48:[35]

Hold your tongue, Heimdallr,	*Þegi, þú, Heimdallr,*
for you in ancient days	*þér var í árdaga*
the ugly life was ordained:	*it lióta líf um lagit:*
with muck on your backside	*aurgo baki*
you'll always be	*þú munt æ vera*
and keep awake as watch for the gods.	*ok vaka vǫrðr goða.*

Insults of this kind are symbolic attacks, and they stretch and distort reality into harmful statements about the victim's nature. In this case, Loki claims that Heimdallr was given, or even forced into, an unpleasant role as guardian in some distant past. What Ursula Dronke translates as "with muck on your backside," is, according to Meulengracht Sørensen, a hint about obscene unmanliness, passive involvement in homosexual activ-

29 Larrington, *The Poetic Edda*, 53.

30 Helgason, *Eddadigte II: Gudedigte*, 14.

31 Byock, *The Prose Edda*, 37; "Hann býr þar er heitir Himinbjǫrg við Bifrǫst. Hann er vǫrðr goða ok sitr þar við himins enda at gæta brúarinnar fyrir bergrisum" (Faulkes, *Gylfaginning*, 25).

32 Byock, *The Prose Edda*, 37; "... ok heyrir blástr hans í alla heima" (Faulkes, *Gylfaginning*, 25-6).

33 Byock, *The Prose Edda*, 72; "En er þessi tíðindi verða þá stendr upp Heimdallr ok blæss ákafliga í Gjallarhorn ok vekr upp ǫll guðin ok eiga þau þing saman" (Faulkes, *Gylfaginning*, 50).

34 Byock, *The Prose Edda*, 36; "mikill ok heilagr" (Faulkes, *Gylfaginning*, 25).

35 Dronke, *Mythological Poems*, 343.

ity – a typical element in Old Norse insults against men.[36] In this case it is used metaphorically, I think, to hint that Heimdallr's role as a guardian is comparable to such a degrading position. Loki presents the guardian role as a lasting state of dishonour, a state which was decreed for him by someone more powerful, perhaps by Óðinn?

In another Eddic poem, *Skírnir's Journey* (*Fǫr Skírnis* or *Skírnismál*), there is also an allusion to the guardian of the gods (stanza 28). Skírnir threatens the giant maid Gerðr with an unpleasant life if she resists his will. Gerðr will be deprived of her intelligence and her freedom, and she becomes more famous "than the watchman with the gods! / Gape from the bars of the gates!"[37] It is not easy to determine exactly how far the parallel between Gerðr and Heimdallr goes. The poet only says that Gerðr will be as well-known as the guardian of the gods, but probably implies a wider resemblance between the guardian role of Heimdallr and Gerðr's dishonourable and unpleasant future. If Heimdallr was once forced into his guardian role, as Loki claims in *Loki's Quarrel*, then perhaps the poet also alludes to that aspect in *Skírnir's Journey*?

Thus, Heimdallr was not only born on the periphery, at edge of the world, as *The Song of Hyndla* states. Due to his miraculous birth by nine women, he is distanced from the destructive powers of kinship which define Óðinn's world order. Within this order he plays a marginal role as a guardian and his residence is located on the periphery, on the borderland against the giants. Thus, it is perhaps after the collapse of Óðinn's world order that Heimdallr's potential is fully realised?

Christianization

The Norwegian historian of religion, Gro Steinsland, has argued that Heimdallr is identical with a powerful being referred to in *The Seeress's Prophecy*.[38] The poem describes the final battle of *ragnarǫk*. The earth has sunk into the sea and once again risen from it. Then the poem narrates that some of the gods have survived or been resurrected and will inhabit the new world, which is described as a more harmonic place than the previous world. Then, one of the manuscripts (*Hauksbók*) presents a ruler character:

> Then the powerful, mighty one, he who rules over everything, *Þá kemr hinn ríki / at regindómi,*

---

36 Preben Meulengracht Sørensen. "Loki's Senna in Ægir's Hall." In *Idee. Gestalt. Geschicte. Festschrift Klaus von See. Studien zur europäischen Kulturtradition*, ed. Gerd Wolfgang Weber (Odense: Odense University Press, 1988), 239-59, at 253.

37 "Víðkunnari þú verðir/en vǫrðr með goðom!/Gapi þú grindom frá!" (Dronke, *Mythological Poems*, 382).

38 Gro Steinsland, *Norrøn religion. Myter, riter, samfunn* (Oslo: Pax Forlag, 2005), 220-27; Steinsland, *Den hellige kongen. Om reiligion og herskermakt fra vikingtid til middelalder* (Oslo: Pax Forlag, 2000), 169-8; Steinsland, *Det hellig bryllup og norrøn kongeideologi. En analyse av hierogami-myten i Skírnismál, Ynglingatal, Háleygjatal og Hyndluljóð* (Oslo: Solum forlag, 1991), 275-303.

will come from above, to the judgement-place of the gods.[39] *ǫflugr, ofan, / sá er ǫllu ræðr.*[40]

Traditionally, "the powerful, mighty one" has been explained as a reference to Christ, as a late Christian addition to the poem. That a powerful being descends from heaven and rules everything certainly indicates Christian influence. However, Steinsland claims that, from a Christian perspective, it might be problematic to integrate Christ into a poem which exclusively presents references to pagan mythology – as a ruler in a world where some of the pagan gods are still active forces. Steinsland would rather see the description of the new and possibly harmonic world, and the ruler of this world, as a pagan transformation of Christian beliefs and motifs.

Steinsland identifies the unnamed ruler as a pagan parallel to Christ, and she claims that Heimdallr was the god who was given this role. She points to the fact that the ruler descends from above (*ofan*) and that Heimdallr's residence is called Himinbjǫrg, which is associated with heaven (*himinn*). In addition to this, there are terms in the stanza which elsewhere are applied in references to Heimdallr, for example *ǫflugr* "powerful," which also is present in the stanza from the *The House Lay* which I mentioned in the introduction.[41] Steinsland also considers Heimdallr's origin the major reason why he became the god of a new world order, but she has mainly focused on the fact that his origin in nine giantesses signifies a unity and a unique kind of female creativity. My contribution to this interpretation is the focus on the difference between female and male lines of descent in Old Norse conceptions of kinship, and the theory that Heimdallr's all-female origin frees him from the fateful consequences of patrilineal kinship.[42]

The most interesting parallel to the stanza in *The Seeress's Prophecy* is to be found in the poem *The Song of Hyndla*, which presents the most extensive description of Heimdallr's birth in the Old Norse sources. After the description of his birth and his powers (stanza 35–38), the poem turns to Loki and describes some of the more perverted actions which he is associated with, for example fathering a wolf and mothering an eight-legged stallion (stanzas 40–41). After the two beings are contrasted, a stanza informs us that the end of the world is closing in. Two characters are then presented, one of them probably Heimdallr (stanza 43):

One was born greater than all, *Varð einn borinn / ǫllum meiri,*
he was empowered with the strength of earth; *sá var aukinn / iarðar*

[39] Larrington, *The Poetic Edda*, 12.

[40] Dronke, *Mythological Poems*, 87.

[41] Steinsland, *Det hellige bryllup*, 295.

[42] Bernt Øyvind Thorvaldsen, *Mǫgr átta mæðra ok einnar: mytane om Heimdallr i lys av førestellingar om slektskap* (Bergen: University of Bergen, 2002).

megni;
he is said to be the wealthiest of princes, *þann kveða stilli / stórauðgastan,*
closely related to all the families.[43] *sif sifiaðan / siǫtom giǫrvǫllom.*[44]

However, there is also mention of a "second" one or "another" one in stanza 44, which might be even more powerful (*enn mátkari*). In Steinsland's opinion, this is the future ruler Heimdallr, the one mentioned in *The Seeress's Prophecy*. The reason why he is described as another one than Heimdallr, or a second one after Heimdallr, is, according to Steinsland, due to the fact that Heimdallr will return as a *transformed* being. Although I find that explanation somewhat pressed, the idea of a pagan "translation" of Christ behind the future rulers described in The *Seeress's Prophecy* and *The Song of Hyndla* seems probable.

In *The House Lay* the poet describes a battle between Loki and Heimdallr over the sea-kidney. The sea-kidney, could in my opinion, refer to the world as an island which rises from the sea after the doom. In that case, we are dealing with a myth in which Heimdallr and Loki fight for future world domination. This reading is in harmony with Steinsland's theory of Heimdallr as the future ruler of the world in *The Seeress's Prophecy*. The wise and famous Heimdallr is – due to his birth from nine mothers – immune to the fateful and destructive consequences of the creation. Loki, however, represents the weaknesses of the old world, since he is the son of a giant and possibly a goddess. Unlike Heimdallr, he carries in him the opposition between gods and giants, the conflict which makes the world a fragile and temporary phenomenon. Heimdallr is – like Christ – something new, something which gives hope of a future state of harmony, an existence which is not based on feuds between gods and giants, but on freedom from such devastating conflicts. When Úlfr describes the battle over the new world in *The House Lay*, his references to the ancestry of Heimdallr and Loki serve to underline the transition from a fragile, faulty world to new and more harmonic existence. If this interpretation is right, it probably represents influence or inspiration from Christianity, which should be expected as late as the 980s.

A couple of decades after Úlfr performed *The House Lay*, another poet, Þorvaldr veili, asked him to kill the Christian missionary Þangbrandr. Úlfr did not consent to the idea and he warned his colleague of the consequences of attacking missionaries. The warning turned out to be justified: Þorvaldr veili composed an insult (*níð*) against Þangbrandr, after which the missionary promptly killed him. Thus, perhaps Úlfr was wise enough to realise that the arrival of Christianity was unavoidable and perhaps he

43 Larrington, *The Poetic Edda*, 259.
44 Jón Helgason, *Eddadigte II: Gudedigte*, 87.

did not mind too much.[45] The second stanza of his *House Lay* indicates that the Christian framework, with its belief in a better and more perfect existence, had already affected the pagan way of thinking when the missionaries arrived in Iceland.[46]

---

45 Bergsveinn Birgisson, *Inn i skaldens sinn: Kognitive, estetiske og historiske skatter i den norrøne skaldediktningen* (Bergen: Universitetet i Bergen, 2008), 418-20.

46 The arguments in this article were partly presented in Norwegian in Bernt Øyvind Thorvaldsen, "Kampen om den kommende verden," *Nordica Bergensia* 29 (2003): 171-93. The journal *Nordica Bergensia* has now ceased to exist.

THE QUALITY OF ISLANDS IN MIDDLE HIGH GERMAN LITERATURE

Gerhard Jaritz

Medieval literature can open up insights into any constructed values and qualities shared by a society or by specific groups in a society. One must be aware of the fact that all these values have to be seen as representations or reflections of positive or negative connotations that cannot, however, be taken as 'realities' of life. This must, in particular, be considered for those aspects of texts and contexts that might be seen as parts of the authors' and the recipients' quotidianity, where the contents of the texts sound especially 'realistic,' as having been taken from the author's and his audience's life.

This is rather different for the situations and objects dealt with by authors which they and also their audience clearly did not know from their own experience. Here it is much clearer that writers created or constructed objects and activities that they had become acquainted with in stories which they had read or heard, copied or translated, in products of their own fantasy but based on such stories, and certainly not by anything that they had experienced themselves.

This can often be seen as the actual background and pre-condition for the situation when authors used islands in their literary product, in particular when they, and also most of their recipients, lived in an area where no islands existed, particularly no islands in the sea. This is true for many Middle High German texts. The authors did not have any personal experience, but offered something which was their own creation or a copy of a construction by someone else from whom they had taken it over or translated it.

The aim of this analysis, therefore, cannot be an effort to reconstruct the 'reality of islands' in the twelfth, thirteenth, fourteenth or fifteenth centuries, but will be an attempt to apply a comparative approach to Middle High German 'literary islands,' their value and qualities and the recurring patterns of the latter's constructions.[1]

1 Of particular help and relevance for this contribution was the 'Middle High German Conceptual Database' (http://mhdbdb.sbg.ac.at:8000/index.en.html), developed since 1992, originally at Bowling Green State University and the University of Kiel (Germany), and since 2002 located and attended at the University of Salzburg in Austria.

Generally, it has to be emphasized that islands in literature are only seldom 'neutral.' "The languages they speak, evoke or suggest, are mysterious, baffling, elusive, deceitful, characterized by uncanny, disturbing sounds and signs, ..."[2] Often, it is not just an island that is dealt with, but an island characterized by specific attributes or qualities. This means that the creation or construction of 'literary islands' was connected and happened in the context of the construction of their quality or value, which strengthened the connotation of the textual environment. One was confronted with a specific, unknown object that was special or, better, was made special in the course of its textual representation, which reflected an appraisal and evaluation of it.[3] A number of different specific attributes and qualities were connected and put into context with islands and the beings living on them. The unknown had to be made known or even familiar and was supposed to become subject to a validation which followed the expectations of the audience and the generally developed patterns of judging and recognizing 'other,' that is, unknown, objects. Many of these 'literary islands' are isolated, distant in the sea, and alien, either evaluated as such generally without further description or explained more accurately and in detail. [4]

In some of these more detailed contexts one can clearly recognize a geographical background or the author's knowledge, a mixture of reality and myth, often going back to ancient times. This can, for instance, be traced for Thule.[5] In the *Lanzelet* of Ulrich von Zatzikhoven, written after 1192, one finds such a description of it. The broad island is full of wonders. In the week before Christmas the days are so short that a messenger is almost unable to run for half a mile before night falls. In the summer the

---

2 Sergio Perosa, *From Islands to Portraits. Four Literary Variations* (Amsterdam: IOS Press, 2000). For a still-valuable study on the representation of islands in the Middle Ages see also William H. Babcock, *Legendary Islands of the Atlantic. A Study in Medieval Geography* (New York: American Geographical Society, 1922, repr. Freeport: Books for Libraries, 1975).

3 Ibidem: "Throughout the course of literature, islands have accumulated uncanny connotations and peculiarities of linguistic definition and expression. They speak strange languages; ... Reading islands is often reading oneself: they are by definition the place where questions are asked, and (tentative) responses are sought."

4 See, e. g., Rudolf von Ems (c. 1200-1254), *Barlaam and Josaphat* [Franz Pfeiffer (ed.), *Barlaam und Josaphat von Rudolf von Ems* (Leipzig: Göschen'sche Verlagshandlung, 1843, repr. Berlin: Walter de Gruyter, 1965), col. 128, v. 5073-76):
dar nâch wart er schiere gesant
in ein vremedez einlant,
dâ er von hungers nôt verdarp,
von vroste und von unrâte erstarb; ibidem, col. 157, v. 6256: ...*verre in dem mer ein einlant.* ...; see also note 10.

5 See, for instance, Germaine Aujac, "L'ile de Thulé, mythe ou réalité," *Athenaeum,* nuova series 66 (1988): 329-43; Donna A. Vinson, "The Western Sea: Atlantic History before Columbus," *The Northern Mariner* X, 3 (2000): 5; Folker Reichert, "Mythische Inseln," in *Burgen, Länder, Orte*, ed. Ulrich Müller and Werner Wunderlich (Konstanz: UVK Verlagsgesellschaft, 2008), 642-3.

days are much longer than here. One will never have heard such strange things as are told about this island.[6]

One can find another, similar description of a distant island in Seifrit's mid-fourteenth-century *Alexander* romance. The emperor wants to find out where the end of the world is. It is a large island where one cannot live and there is neither night nor day, because of the cold there. The place is called *Polus antarcticus*.[7]

Distant islands were sometimes connected with savagery and wildness.[8] The attribute and quality 'wild' is common for the characterization of such 'literary islands' and the beings living on them. Often the isles themselves are just called wild, without

---

6 Wolfgang Spiewok (ed.), *Ulrich von Zatzikhoven, Lanzelet* (Greifswald: Reineke-Verlag, 1997), 313-4, v. 7994-8007:
... daz Thîle ist ein einlant,
ein breit insele in dem mer.
dâ ist von wunder manic her,
diu nieman kunde geahten.
ein wochen vor wîhnahten
sint sô kurz dâ die tage
nâch Rômaere buoche sage,
dâ manic wunder an stât,
daz ein loufer kûme gât
vor naht ein halbe mîle.
die tage sint ouch ze Thîle
ze sumer langer danne hie.
ir envrieschent vremder maere nie
dan uns dannen sint geseit.

7 Paul Gereke (ed.), *Seifrits Alexander aus der Strassburger Handschrift* (Berlin: Weidmann, 1932, repr. Hildesheim: Weidmann, 2005), 102, v. 6166-79:
... er muest ersuechen und ervinden
fuer war die rechten mer
wo der welt endt wer.
do cham er nachant dar
in ain grossew insula,
do der welt ende was,
gegen ainem tall (gelaubet das!)
da man furbas nacht noch tag
vor chelten nit gewonen mag.
die selb stat nent man sus
Polus antarticus.
darinne man ain veste stat
vor maniger zeit gemauret hat.

8 Perosa, *From Islands to Portraits*, 4: "All these medieval islands are desert, hostile, savage places; they overwhelm voyagers and shipwrecked people with horrors; they provide the seat for tests of solitude and alienation, for rites of passage."

any further or detailed characterization.[9] In the so-called *Prose-Lancelot*, the German translation and adaptation of a French trilogy from 1215-30 that was begun about 1250, one comes across an island that is very wild and hidden, which can be seen as a wonder. The same text leads the reader to other distant islands, some lacking humans: "They came to distant islands far away from people where they did not find anything else but wild beasts and where they had many adventures which they survived with the help of the Holy Spirit."[10] A number of times this savagery and wildness is described explicitly in great detail, mainly by using wild creatures. In the *Prose-Lancelot*, Parceval came to an island not knowing where he was. He wanted to be informed about the island's name but did not know how to find out, as there was no town, no castle; there were only wild animals around him: bears, leopards, and dragons.[11]

Seifrit's Alexander comes to a large island populated with creatures from the edges of the world, which are also known from other evidence[12]: people without heads having the mouth, eyes and ears on their chest, being twelve feet long and having a waist measure of seven feet, *gelaubet das* ("just believe that"), with a body of golden col-

---

9 See, e.g., Rudolf von Ems, *Alexander* [Victor Junk (ed.), *Rudolf von Ems, Alexander. Ein höfischer Versroman des 13.* Jahrhunderts, vol. 1 (Leipzig: Verlag Karl W. Hiersemann, 1928, repr. Darmstadt: Wissenschaftliche Buchgesellschaft, 1970), 186, v. 5205-06]:
… vil manec wildez einlant,
der name ist uns vil unbekant …; ibidem, vol. 2, 600, v. 17390-94:
Tiutschland und Galliâ
wirt (von den Nachkommen Ismahels) *mit urliug an geritn,*
ouch werdent Rômaer nicht gemitn,
die verjagt daz selbe her
in wild inseln in dem mer, …; Konrad von Würzburg (c. 1230-1287), *Der Trojanische Krieg* [Adelbert von Keller (ed.), *Der Trojanische Krieg von Konrad von Würzburg* (Stuttgart: Litterarischer Verein, 1858), 208, v. 17492]: … *in manic wildez einlant* …; ibidem 586, v. 49088: … *in manic insel wildeclich* …; ibidem, v. 49116: … *in manic insel wilde* …; Reinhold Kluge (ed.), *Lancelot. Nach der Heidelberger Pergamenthandschrift Pal. Germ. 147*, vol. 3 (Berlin: Akademie-Verlag, 1974), 274: … *und da gelenten sie zwuschen zweyn leyen an einer inselen, die was sere wild und als verborgen das es wunder was;* Heinrich von dem Türlin, *Diu Crone* (c. 1230) [Gottlob Heinrich Friedrich Scholl (ed.), *Diu Crone von Heinrich von Türlin* (Stuttgart: Litterarischer Verein, 1852, repr. Amsterdam: Rodopi, 1966), 63, v. 5101]: … *in ein wildez einlant* …

10 Kluge, *Lancelot*, vol. 3, 344: … *und sere dicke gelenten sie an fremden inselen und ferre von lúten, da sie anders nit funden dann wilde tier, da sie manch wúnderlich abentur funden der sie zu ende kamen so mit yrer biederbkeyt, so mit der gnade des heiligen geystes, der yn allczyt halff,* …

11 Ibidem, 124-5: *da bruofft er das er were gefürt in ein insele, das er nit wüste wo und hett es gern gewüst, dann er kuont nit wissen wie er es möcht erfarn, wann er sach wedder vesten noch burg, da ymandt möchte inne gesin als yn ducht, und noch dann, so was er nit alleyn, er sehe umb yn wilde tiere, beren, leoparten und trachen. und da er sich in solcher stat sahe, da was er nit wol zugemach, wann er forcht die wünderlichen tiere, das sie yne nit mit gemache ließen, als yn ducht, und yn döten, ob er sich nit konde geweren.*

12 See, e. g., Gerhard Jaritz, "From the Peripheries to the Centres and back again: Visual Culture and the Edges of this World," in *The Edges of the Medieval World*, ed. idem and Juhan Kreem (Budapest and New York: Central European University Press, 2009), 21-38.

our.[13] In a large forest they saw animals that had bodies like horses and feet like a lion's. They were thirty feet high and had a waist measure of twelve feet; their voice was "unsweet" and they had a terrifying gaze.[14] As above, it was sometimes indispensable to emphasize that the offered story was true.[15] Wildness and the marvels of islands could clearly lead to the situation that no one would believe the stories.[16]

---

13 Gereke, *Seifrits Alexander*, 112, v. 6829-42:
do er den sig alda gewan,
do hueb er sich zu hant von dann
und cham zu ainem grossen wag,
darinn ain grassew insula lag.
da sachens leut inne gan,
die warn all des hauppts an,
der mund, augen und die orn
stuenden in an der prust vorn.
ir leng zwelef fuess was,
die dikch suben, (gelaubet das!).
sy warn ungeschaffen gar
und an dem leib goldvar.
sy sachens durich ain wunder an
sy tetten in nit und liessens gan.

14 Ibidem, 112, v. 6845-56:
sy chomen fur einen grassen wald,
da sachens tier ungestalt.
sy warn dikch und gros,
der leib geschaffen als ain ros,
diē fuess geleich ainem leben.
sie lieffen vor in und en neben.
ir hoch was dreyssig fuess,
ir stym was gar unsuess,
sy warn zweleff fuess dikch,
ir augen betten frayslich plikch.
man sach sy vrayslich geparn,
yedoch ir nit vil warn.

15 E. g., Junk, *Rudolf von Ems, Alexander*, vol. 2, 626, v.18156-62:
ein insel nâhe bî uns lît,
dâ wonet inne, daz ist wâr,
hundert tûsent megede klâr,
die sint sô minneclich gevar
daz nie wîbes lîp gebar
als gar wunschlîchiu kint
sô die selben megede sint.

16 Reinhold Kluge (ed.), *Lancelot. Nach der Kölner Papierhandschrift W. f° 46* Blankenheim und der Heidelberger Pergamenthandschrift Pal. Germ. 147*, vol. 2 (Berlin: Akademie-Verlag, 1963), 698-9: … *also das ich qwam an die wunderlich insel … und ist so vil wunders inn der insel das syn nymands gleuben mocht, er wer dann da gewest, wann es ist der grost gewalt dainn von allen zeuberyen. und kem der best ritter der welt da hien, er fund synen glichen. dann es mocht kein so abenturlich ritter da hien komen noch keyner von so großer sterck dar komen, er fund abentur daselbst wie er sie haben wolt.*

One could find an island with a terrifying dragon on it who spit fire out of its mouth,[17] or another one with a bear;[18] and there was an island with the largest and most miraculous giant of this world living on it, who shouted "You must be dead!"[19] There was also an island where sirens lived, as Konrad von Würzburg reflected in his mid-thirteenth-century *Trojan War*.[20]

One can also find isles with only women living on them: in Seifrit's *Alexander*, for instance, there is such a large island with a queen.[21] Similar situations are described

---

17 *Dietrichs Flucht* (Dietrich's Flight), end 13th c. [Ernst Martin (ed.), *Alpharts Tod, Dietrichs Flucht, Rabenschlacht* (Berlin: Weidmann, 1866, repr. Berlin: Weidmann, 1975), 80, v. 1536-38 and 1544-51]:
... dô kom ein sturm, der tet in wê
und sluoc si leider alzehant
ûz in ein einlant. ...
in der zît dô lief dort her
ein wurm ungehiuwer,
dem vuor wildez viuwer
ûz ze sînem munde,
swenne er blâsen begunde.
sîn stimme unmaezliche erdôz.
ez was ein tier kûm alsô grôz
sam in der mâze ein serpant.

18 Konrad von Würzburg, *Partonopier and Meliur* [Karl Bartsch (ed.), *Konrad von Würzburg, Partonopier und Meliur* (Vienna: Wilhelm Braumüller, 1871, repr. Berlin: Walter de Gruyter, 1970), 262, v. 18163-73]:
... daz in den hof ein bote kam
für den keiser lobesam
und im dâ seite maere,
daz ein insel waere
dâ bî gelegen in dem mer,
ûz der striche ân alle wer
in daz lant ein michel ber,
der schüefe dâ nâch sîner ger
der werlte schedelîche not.
vih und liute waeren tôt
von sîner grimmekeite ...

19 Kluge, *Lancelot*, vol. 3, 283: ... *da fand er off einer insele ein riesen, den aller grösten und den aller wünderlichsten von der welt, der yne anschrey: 'du müst dot syn!' und dieser forcht sich zu sterben, und er sah umb sich, wann er sah nirgents nichts, da mit er sich mocht beschütten.* ... See also the reference to a giant on an island in Scholl, *Diu Crone*, 68, v. 5520-22:
Assiles was der ris genant
vnd saz in einem einlant,
daz was starch wilde. ...

20 Keller, *Der Trojanische Krieg*, 586, v. 49096-99:
... hier under sô gesuocte sich daz,
daz in warf ein starker wint
in ein insul, dâ eht sint
die Sîrênen wonende.

21 Gereke, *Seifrits Alexander*, 81, v. 4896 and 4901-02:

for the island of Lemnos in *The Trojan War*[22] and for the island of Nio in the songs of Oswald of Wolkenstein (c. 1377-1445).[23]

The particularity of 'literary islands' is not only confined to the faraway distance, their savagery, wildness, marvels, and adventures. Their peculiarity can also refer to exceptionally positive factors, that is, for instance, the beauty of the island, its wealth and riches.[24] Such an aspect of operating with opposites is generally rather typical for medieval texts in which contrasts play an important role, that is, positive qualities may generally exist and be relevant at the same time as negative connotations.[25] The *Alexander* romance of the Bohemian Ulrich von Etzenbach from the second half of the thirteenth century leads Alexander to an island on which there was a town incomparable in

---

es wer ein grossew insulla,...
da sind nicht wann frawen innen,
die habent selb ain chuniginne ... See also note 15.

22 Keller, *Der Trojanische Krieg*, 165, v. 13816-23:
Lemnos ein insel ist genant,
dâ niht wan frouwen inne sint,
dar în sol ich mîn liebez kint
nû füeren unde bringen.
im kann niht misselingen
bî wunneclichen wîben
diu zuo der manne lîben
tragent weder sin, noch ger ...

23 Karl Kurt Klein (ed.), *Die Lieder Oswalds von Wolkenstein* (Tübingen: Max Niemeyer Verlag, 1962), 51-2, song 18, stanza VI, v. 1-4:
vierhundert weib und mer an aller manne zal,
vand ich ze Nio, die wonten in der insell smal;
kain schöner pild besach nie mensch in ainem sal,
noch mocht ir kaine disem weib geharmen.

24 See, e. g., Junk, *Rudolf von Ems, Alexander*, vol. 1., 144, v. 3981-84:
als er daz lant an sich genam,
dô kêrt er in Hostîam.
daz was ein rîchiu insel wît,
die twanc er an der selben zît. ...; ibidem, 302, v. 8653: ...*diz wâren rîchiu einlant* ...; *The Younger Titurel* (c. 1272) [Kurt Nyholm, (ed.), *Albrechts Jüngerer Titurel*, vol. III/2 (Berlin: Akademie Verlag, 1992), 302, stanza 5598]: ... *vil manic insel riche* ...; Konrad von Würzburg, *Engelhard* [Ingo Reiffenstein (ed.), *Konrad von Würzburg, Engelhard* (Tübingen: Niemeyer, 1982), 197, v. 5234-37]:
... *alsam ein irdesch paradîs*
beschoenet stuont diz einlant.
swem ie gezierde wart bekant,
der möhte ez balde gesten.

25 Cf., e. g., Gerhard Jaritz, "Der Alltag der Kontraste. Muster von Argumentation und Perzeption im Spätmittelalter," in *Kontraste im Alltag des Mittelalters*, ed. idem (Vienna: Verlag der Österreichischen Akademie der Wissenschaften, 2000), 9-23.

its beauty, strength, and wealth. No other town could resemble it.[26] In Seifrit's *Alexander* one came to an island on which a beautiful temple stood, where they met a beautiful stag.[27]

* * *

Let me summarize: Middle High German 'literary islands' are not neutral. They were determined and characterized by the construction of their attributes, qualities and values. The latter made these islands

- distant and isolated;
- wild and dangerous;
- marvelous; or
- rich and beautiful.

As islands could be something distant, unknown and particular for the authors and their audiences, it was more or less indispensable to use such a characterization. Sometimes it seems that it did not matter if the characterization went into a positive or negative validation: The most important role of medieval 'literary islands' seems to have been that they were something extraordinary to be made more familiar with the help of creating special qualities that one already understood or could imagine. The quasi-knowledge, impression and evaluation of islands by members of medieval society who did not know

---

26 Wendelin Toischer (ed.), *Ulrich von Eschenbach, Alexander* (Tübingen: Litterarischer Verein, 1888, repr. Hildesheim and New York: Olms, 1974), 611, v. 22954-58:
... er zogt dâ er ein insulen vant,
dar inne ligen ein stat:
waz er ie stete gesehen hat
der keine wâren der gelîch:
schoene, veste unde rîch ...
sie was in wunnen schouwe. See also *Buch von Troja* (1392) [Hans-Josef Dreckmann (ed.), *Das "Buch von Troja" von Hans Mair. Kritische Textausgabe und Untersuchung* (Munich: Fink,), 14]: *in der insel Colcos waz zu den zeiten diu haptstat dez landes, diu hiezz Jaconites, ain grozziu mehtigiu stat, mit schönen muren, und hochen turn usswendig wol geziert. und innerhalb waz diu stat mit schönen balasten und hüsern, dar inn manig gut ritter und über die mauss vil grozz volks wandelt.* , ...

27 Gereke, *Seifrits Alexander*, 25, v. 1397, 1399-1402, 1405-08:
...und chertten gegen ainer insula, ...
die was gepaut schon
irem werden gott Amon
ain schoner tempel dar inne
zu lob und zu mynne. ...
do sy zogten under wegen
do sachen sy das in engegen
gelauffen cham ain schoner hiers,
ich wais von wann, gelaubt mirs!

any of them personally were certainly influenced by such constructions and their dissemination.

SEASONAL ISOLATION IN THE COMMUNICATION IN LIVONIA[1]

Juhan Kreem

Being isolated seems to be in the nature of every island. Isolation creates islands. On the other hand, the title of our workshop, "Isolated Islands," when not pleonastic, implies that there are islands which are isolated and there are others which are not. In the medieval communication network on the Baltic some islands, rather than being isolated backwaters, were crossroads of international trade. The most prominent example among them is Gotland. Speaking in more general terms, Livonia, which is geographically part of the continent, was an island overseas from the Hanseatic perspective or from the perspective of the Holy Roman Empire. This image of an island can be seen most notably in a historiographic legend of the sixteenth century, according to which the merchants of Bremen sailed out and discovered Livonia (the so-called *Aufsegelung*).[2] True, a land-route existed between Livonia and the West, but it always remained insecure because of the Lithuanians, and the bulk of trade went over the seas.[3] The history of Hanseatic communications in general, and maritime history in particular, are about connections and communication, tending to ignore the isolation side of the story. One may ask then: Is there isolation in the history of communication?

There is, of course. But when one speaks of isolation then it is reasonable to ask what kind of features it has. Isolation, in a most general way, is the lack of connections. There are places which have good connections and others which do not, but it is possible somehow to get everywhere humans live. It is possible, but it takes time, and

---

1 The research was done under the auspices of Estonian Science Foundation grant no. 7129.

2 Paul Johansen, "Die Legende von der Aufsegelung Livlands durch Bremer Kaufleute," in *Europa und Übersee: Festschrift für Egmont Zechlin*, ed. Otto Brunner (Hamburg: Hans Bredow Inst., 1961), 42-68.

3 On the Hanseatic communication networks see: Friedrich Bruns and Hugo Weczerka, *Hansische Handelstrassen,* Textband (Cologne: Böhlau, 1967), Atlasband (Cologne: Böhlau, 1962); Jörgen Bracker (ed.), *Die Hanse, Lebenswirklichkeit und Mythos: eine Ausstellung des Museums für Hamburgische Geschichte in Verbindung mit der Vereins- und Westbank.*, vol. 1 (Hamburg: Museum für Hamburgische Geschichte, 1989), chapter "Schiffahrt und Verkehr," 561-607; Raoul Zühlke, "Der Verkehr im Umfeld der Oberzentren Bremen und Riga um 1300 im Vergleich – die Auswirkungen naturräumlicher Bedingungen," *Zeitschrift für Ostmitteleuropa-Forschung* 53 (2004): 475-503.

this is a crucial factor in discussing isolation. Furthermore, the time factor brings communication in pieces, when a caravan or fleet arrives or when a letter arrives from Rome. These pieces of communication are followed by days, weeks, or months of silence. The dynamics are not only caused by human agency (the will to get somewhere, talk to someone), but also by external conditions. There are periods, which we call seasons, when more communication takes place. My aim in the following is to sketch the seasons of communication and isolation on the Baltic landscape.

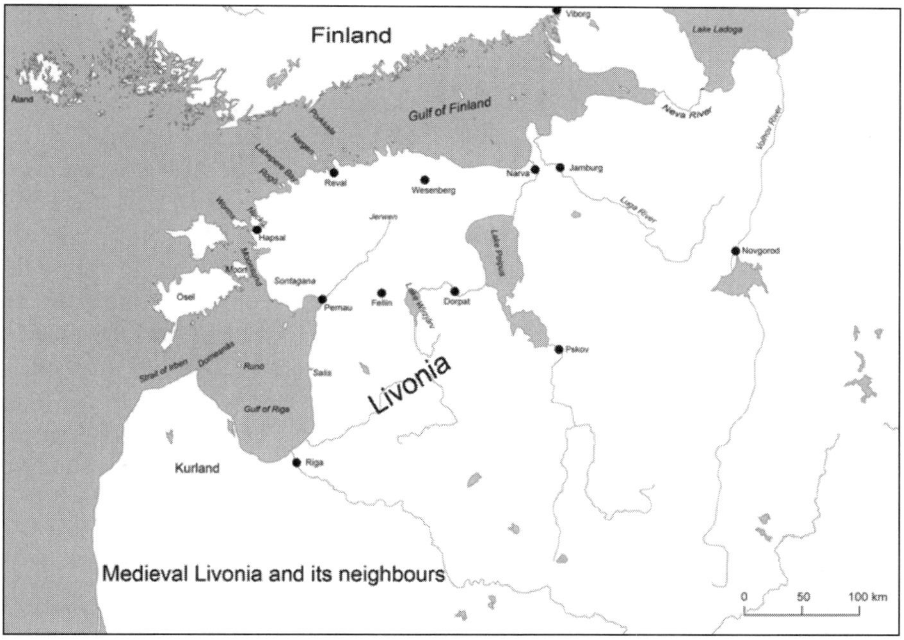

Fig. 1

Water is crucial in the communication system of the eastern Baltic. Primarily the sea, but also the great rivers provide ways of communication. The coastal towns owe their wealth to overseas trade. Water transport has been used as the most important explanation in discussions of the topic of why the towns appeared where they did. While a location near the estuary of a river is a good explanation for Riga, Narva, and Pernau, the situation of Reval has been connected to the fact that it is close to the place where the ancient waterway crossed the Gulf of Finland at the narrowest place between (Porkala and Nargen).[4] Most recently, Anton Pärn has connected the location of Hapsal with the waterway through the straits of Nuckö and Worms.[5]

Water is not always navigable, however, as everywhere else in the world there are ill winds and storms. Stormy winters also define the navigational season, for example, on the Mediterranean.[6] In the Baltic the freezing of the sea limits the possibilities for shipping even more clearly. Not only the ice cover itself, but the ice edge which cuts through the planking of wooden ships or the freezing of ships above the waterline (which makes them capsize) were dangerous. Even nowadays, despite climatic change, ice seriously influences navigation on the Baltic Sea.

The navigable season is reflected in many sources.[7] The thirteenth-century chronicle of Henry of Livonia hints in this direction: the crusaders arrive and leave on ships, which clearly has seasonal dynamics.[8] There are, of course, more detailed, sober, and exact sources on the navigation seasons. Gunnar Mickwitz has established the average navigation period as from April to November-December according to sixteenth-century account books of Revalian merchants.[9]

Hanseatic legislation established navigational seasons. According to thirteenth-century Hamburg maritime law, after St. Martin's day (11 November) ships must remain in winter quarters.[10] More detailed regulations from the beginning of the fifteenth century fixed the navigation season from the Cathedra Petri (22 February) to St. Martin's Day. Exceptions were made for cargoes of beer and herring, which could sail from Candelmas (2 February) to St. Nicholas' Day (6 December).[11] Fixing the exact dates for the navigational season must not be taken as evidence for a history of climate. The

---

4 Leo Tiik, "Keskaegsest mereliiklusest Balti merel ja Soome lahel" (On the medieval sea-traffic in the Baltic Sea and Gulf of Finland), *Eesti Geograafia Seltsi Aastaraamat 1957* (Tallinn 1957), 207-30, here 210-2.

5 Anton Pärn, "Die Rolle der Wasserstraße bei der Ortswahl der Stadt Haapsalu (Hapsal)," in *Lübeck Style? Novgorod Style? Baltic Rim Central Places as Arenas for Cultural Encounters and Urbanisation 1100-1400 AD*, ed. Muntis Auns (Riga: Nordik, 2001), 97-108.

6 Fernand Braudel, *The Mediterranean and the Mediterranenan World in the Age of Philip II*, trans. Siân Reynolds (London: Harper Collins, 1976), 231-72.

7 Specialized studies on the history of climate are: Andres Tarand and P. Øyvind Nordli, "The Tallinn Temperature Series Reconstructed Back Half a Millennium by Use of Proxy Data," *Climatic Change* 48 (2001): 189-99; Andres Tarand and Tiit Kallaste (ed.), *Country Case Study on Climate Change Impacts and Adaptation Assessments in the Republic of Estonia* (Tallinn: SEI, 1998), 15-21.

8 E. g., Leonid Arbusow and Albert Bauer (ed.), *Heinrici Chronicon Livoniae* (Darmstadt: Wissenschaftliche Buchgesellschaft, 1959), (hereafter: *HCL*), VII, 1; VIII, 1; XI 1; XII 1; etc. Summary of the data on shipping in the chronicle see Friedrich Benninghoven, *Rigas Entstehung und der frühansische Kaufmann* (Hamburg: August Friedrich Velmede, 1961), 39-41.

9 Gunnar Mickwitz, *Aus Revalar Handelsebüchern. Zur Technik des Ostseehandels in der ersten Hälfte des 16. Jahrhunderts* (Helsinki: Societas Scientarum Fennica, 1938), Table no. 12, 13 on156.

10 Rolf Sprandel (ed.), *Quellen zur Hanse-Geschichte* (Darmstadt: Wissenschaftliche Buchgesellschaft, 1982), 400, § 13.

11 Ibidem, 418, § 26, 27. 24.

northern part of the Baltic Sea was most likely still covered with ice in February. These regulations were more connected with the Hanseatic will to level the commercial environment for the members of the league, to cut off risky profits from sailing before the others. There were even occasions when the start of the navigational season was postponed to Pentecost for political reasons.[12]

There is plenty of evidence on sailing in extreme conditions and on cases when a ship could not reach its destination because of the ice. In the autumn of 1204, when the crusaders returned from Livonia to Germany, they were trapped by ice somewhere near the coast of Denmark on 2 November. As was often the case, the ship was left in the ice and the crusaders went over it to the mainland.[13] The chronicles record many instances of the freezing in of the fleet; one of the most famous and adventurous is that in the *Lübeck Chronicle* of Reimar Kock. It narrates how the fleet of Wismar, which was sent to the relief of Stockholm in 1394, stuck in the ice and had to fight off attacks of the Danes, first by making ice-slippery barricades, then by cutting trenches into the ice, which froze during the night but broke under the Danes in the morning.[14]

One feature which needs to be stressed when analyzing the structure of the communication network in the Baltic is that the duration of ice-cover varies considerably in the different regions of the Baltic Sea. Looking at the dynamics of the limit of the ice cover on modern maps, one sees how much faster the straits between the islands freeze than the open coast.[15] This means that when Hapsal or Pernau are already frozen in, Reval might still be open. On the Gulf of Finland the freezing limit not only moves on a north-south axis, but also an east-west one. For example, navigation to Narva is normally hindered much longer by the ice than that to Reval. In Revalian customs accounts (*Schiffslisten*) it is recorded that on 11 April 1440, when the fleet arrived, it could not reach Reval because the bay was still iced in. The fleet stayed in the bay of Lahepere (*achter Ladwenpe*), over 30 km west of Reval, and only arrived in Reval five days later.[16] As in every history of weather there are exceptions. In 1568, a German ship arrived in Reval on 2 February, which the chronicler Balthasar Russow found sensational enough to record in his book.[17]

---

12 Paul Johansen and Heinz von zur Mühlen, *Deutsch und Undeutsch im mittelalterlichen und frühneuzeitlichen Reval* (Cologne: Böhlau, 1973), 442-3.

13 *HCL*, VIII, 3.

14 Ferdinand Heinrich Grautoff (ed.), *Chronik des Franciscaner Lesemeisters Detmar, nach der Urschrift und mit Ergänzungen aus anderen Chroniken*, part 1 (Hamburg: Fr. Perthes, 1829), 495-7: excerpts from the chronicle of Reimar Kock.

15 See, e. g., the homepage of the Finnish portal on Baltic Sea (http://www.itameriportaali.fi/en/tietoa/jaa/jaatalvi/en_GB/jaatalvi/) accessed March, 2010.

16 Reinhard Vogelsang (ed.), *Revaler Schiffslisten, 1425–1471 und 1479–1496* (Cologne: Böhlau, 1992), 318-9.

17 Balthasar Russow, *Chronica der Prouintz Lyfflandt* Scriptores rerum Livonicarum, vol. 2 (Riga and Leipzig: Eduard Frantzen, 1853), 75.

To cut a long story short, ice covers the rivers and sea and stops the navigation of ships. But does it mean isolation? In a way, yes; navigation stopped, but communication did not stop, it took other forms. It is time to turn to the topic of ice as an opportunity.

The most famous episodes on the opportunities of ice are known from military contexts.[18] According to Henry of Livonia, the people of this region had the habit of conducting their military campaigns during Lent.[19] Descriptions of German raids to northern and western Estonia often mention the frozen sea.[20] On the first raid to Revalia in 1219 a route was taken over the frozen sea from Salis to Sontagana; on the way back the crusader army took, according to Henry, an even longer route and spent 10 days crossing the ice.[21] When narrating a ride to Jerwen in 1212, Henry stresses the precondition that the road was frozen.[22] This time the crusaders returned from Jerwen over the ice of Lake Witzjärv.

Ice provided not only a communication route, but also a battlefield. Besides the famous Battle on the Ice in 1242 (on Lake Peipus), many less famous but more important battles were fought in different places. One of them, for example, was fought on the ice of Moonsund in 1270,[23] where the master of the Teutonic Knights in Livonia fell; another on 2 February 1348 in Lithuania on the Sreva River, where the Lithuanian Duke Algirdas fled and the ice failed under him.[24] Winter facilitated ice roads not only over the sea and rivers, but also over swamps. Describing Livonian raids in the fourteenth century, a chronicler, Johann Renner, frequently uses the argument that because of a mild winter the ground was not frozen and therefore the raid on Lithuania failed.[25]

Henry's chronicle seems to imply that the crusaders' campaigns against the western islands of Estonia, Ösel, and Moon could only be conducted in winter. Henry vividly depicts crossing Moonsund over the ice in 1227.[26] Nine years earlier the planned campaign against the Osilians had failed because of a rainy winter; the crusaders had to confine themselves to plundering on the mainland.[27] In 1343, after the rebellion of St. George's night was suppressed on the mainland of Estonia, the campaign against the

---

18 See Friedrich Benninghoven, "Zur Technik spätmittelalterlicher Feldzüge im Ostbaltikum," *Zeitschrift für Ostforschung* 19 (1970): 631-51.

19 *HCL*, IX, 1.

20 *HCL*, XVIII, 5; XIX, 8.

21 *HCL*, XXII, 9.

22 *HCL*, XV, 7.

23 Leo Meyer (ed.), *Livländische Reimchronik* (hereafter: *RC*) (Paderborn: Ferdinand Schöningh, 1876), verses 7847-944.

24 Richard Hausmann and Konstantin Höhlbaum (ed.), *Johann Renner's Livländische Historien* (Göttingen: Vandenhoeck & Ruprecht, 1876), 96.

25 *Johann Renner*, 78, 79.

26 *HCL*, XXX, 3.

27 *HCL*, XXI, 5. dated to Lent in 1218.

Osilians was postponed because the strait was not yet frozen. Only when the ice was hard enough could the Moonsund be crossed and the island pacified.[28] Ösel could be reached not only over Moonsund, but also over the strait of Irben, as was done by the Lithuanians, who raided the Maritima and Livonia through Kurland and Ösel in the winter of 1218-1219.[29] To put it briefly, isolated islands lose their isolation in winter.

The military use of the winter roads is, of course, most visible. But how did winter affect other types of communication? Winter played a prominent role in Hanseatic trade. The Hanseatic treaties with Novgorod distinguish between the summer and winter seasons.[30] In the statutes of the Hanseatic settlement in Novgorod, besides the waterway (*waterwech*), the sledgeway (*sledenwech*) is mentioned.[31] Comparing the prices of transportation between Reval and Narva in winter with sledges or in summer with ships, Mickwitz came to the conclusion that the price of maritime transport was about the same as land transport in winter,[32] whereas land transport in summer was twice as expensive as winter transport.[33] Some other features influenced the price, too, but the general trend shows that winter transport was cost effective. Although it is hard to assess the quantitative importance of the seasons it is quite clear that travel and transportation of goods in winter was not uncommon.

Snow and ice enabled the use of other tracks than in summer.[34] If the road network is a problematic issue in landscape studies, archaeology, and history, winter roads are even more so. A road which is seasonal does not develop a permanent support network (villages), not to mention melting away every spring. The most apparent seasonal variation in the communication is between Reval and Narva. Maritime transport was possible in summer, but in winter goods were transported on land. The track from Dorpat to Narva or Pskov was more consistent; mostly water transport was used in summer and the winter roads followed almost the same pattern. A Flemish nobleman, Ghilbert de Lannoy, describes traveling together with merchants on a frozen river and over Lake Peipus from Pskov to Dorpat in December 1413.[35]

Evidence from later centuries can be found on the communication between two small towns, Pernau and Fellin, separated by great areas of swamp. Pernau, as an

28 *Johann Renner*, 92-3.

29 *RC*, verses 1424-1552, dated to the winter of 1218-1219.

30 Leopold Karl Goetz, *Deutsch-Russische Handelsverträge des Mittelalters* (Hamburg: L. Friedrichsen & Co., 1916), 81, 97, 116.

31 Wolfgang Schlüter (ed.), *Die Nowgoroder Schra in sieben Fassungen vom XIII. bis XVII. Jahrhundert* (Dorpat: Mattiesen, 1911), Wortregister.

32 Mickwitz, *Aus Revaler Handelsbüchern*, 150.

33 Ibidem, 146.

34 Ibidem, 143, without any further elaboration.

35 Theodor Hirsch, Max Töppen, and Ernst Strehlke (ed.), *Scriptores Rerum Prussicarum*, vol. 3 (Leipzig: S. Hirzel, 1866), 443-52, here 447.

exporting port, had part of its hinterland around Fellin; the land route normally had to circle these areas, but in winter the road could be taken straight over the wetlands. These roads are known from the folk tradition and from preserved inns in the middle of the marshes, which are very isolated in summer but had their season in winter.[36] Winter roads were important enough to be depicted as early as the late 18th-century maps of Ludwig August Graf Mellin.[37]

To return to isolation, while summer and winter are each favorable to communication and connections in their own way, do we have isolation at all? In the Baltic world the times of real isolation seem to have been spring and autumn, as is apparent from the perspective of land transport. In the era of natural roads, rainy seasons caused the roads to become impassable. Depending on the year, there was a longer or shorter period between the last/first waterway and first/last sledgeway, i.e., a time without roads. The distribution of military campaigns over the year at the beginning of the 13th century shows that the quietest times were April, October, November and December.[38] Other types of communication are also slow in early spring. In the correspondence of Livonian estates, the *Landtag* is often planned "with the first grass,"[39] which not only implies dry roads, but also better feed for the horses.

On the islands the time of the ice breakup is even more dramatic. In the straits between islands, the time when the ice can no longer carry weight but is still too thick to let ships pass seems to have been the real time of isolation. One of the most vivid descriptions of this season originated in the year 1458, when the envoys of the elected Bishop of Ösel justified their delayed arrival in Rome. In this writing the months of March, April, and May are (*cum grano salis*) described as months when navigation is impossible and men and horses cannot cross the Moonsund. Although the envoys of the bishop tried to cross the ice, they could not reach the mainland and were forced to stay for a considerable time on Schildau Island in the straits. Their subsequent sailing through Moonsund was further delayed because of ice.[40]

---

36 Ilmar Läänelaid and Jaagup Loosalu, "Kaubandus ja unustatud taliteed" (Trade and forgotten winter-roads)," *Eesti Loodus* 11 (1978): 716-21; Ilmar Kask, "Miks Pärnusse otse ei saa?" (Why is there no straight way to Pärnu?), *Eesti Loodus* 4 (1990): 254-5; Jaan Joandi, "Taliteid Viljandimaa ja Pärnu vahel" (Winter roads between Viljandi county and Pärnu), *Eesti Loodus* 4 (1990): 257-61.

37 Eckhard Jäger and Otto Bong (ed.), *Der Atlas von Livland des Ludwig August Graf Mellin [1798]* (Lüneburg: NOKW, 1972), "Der Pernausche Kreis," "Der Fellinsche Kreis."

38 For an analysis according to the *HCL*, see Evalds Mugurevitcš, "Svarigakie celi libiešu un latgala teritorija" (Major communications on the territory of Livs and Latgals), *Arheologija un Etnografija* 3 (1961): 61-81, here 65.

39 Leonid Arbusow (ed.), *Akten und Rezesse der livländischen Ständetage*, vol. 3 (1494-1535), (Riga: J. Deubner, 1910), no. 190, 301#59: *mit dem ersten grasze.*

40 Philipp Schwartz (ed.), *Liv-, est- und kurländisches Urkundenbuch*, vol. 11 (1450-1459) (Riga: J. Deubner, 1905), no. 774.

Spring was a difficult time, but it does not mean that there were no solutions for this season. The seal-hunters of Runö did most of their hunting on the Gulf of Riga exactly in the late winter and early spring months, when the ice was fragile. Adapted to these conditions, their boats had also sledges, so that besides floating they could also be used on the ice.[41] Sledge-boats (*eistuk)* were used also on the northern coast of Estonia.[42]

Weather conditions in the eastern Baltic while stopping one type of communication open up other communication possibilities. The communication pattern of a culture adapts to these conditions and takes advantage of it. Islands could be reached more easily over the ice; marshes which isolated during the summer provided a direct connection in winter. Trade shifted from ships to sledges and back in the course of the yearly cycle. Only the periods of freezing and ice breakup radically limited the communication opportunities among the islands and on the mainland; there were technical solutions for quite extreme conditions. Isolation, as was suspected at the beginning of this paper, is not an absolute category.

---

41 Jorma Friberg, "Paadilugu. Eestirootslaste aerude ja tuule jõul liikuvatest paatidest" (The Boat-story: On Estonian Swedes' rowing and sailing boats), *Akadeemia* 20 (2008): 786-99, 1049-91, here 1065-6. This article also has a bibliography with further readings in Swedish on the issue.

42 Vello Mäss, *Muistsed laevad, iidsed paadid* (Ancient ships, ancient boats) (Tallinn: Horisont, 1996), 150, 155.

UTSTEIN MONASTERY:
AN ISLAND ON AN ISLAND – OR NOT?

Torstein Jørgensen

Introduction

In spite of today's sub-sea tunnels and bridges connecting today's Klosterøy with the Norwegian mainland, there is no doubt that Utstein Monastery is geographically situated on an island. But what did this kind of location mean for the community who once established themselves in this monastic institution? Did they seek this place as an escape from the current and busy life of the town and of society at large? Did the people who settled at Utstein belong to the tradition of monastic groups who sought some remote refuge in the wilderness for the sake of quietness and isolation? What kind of a geographical place was Utstein in the history before and during the centuries of community life here? What links did the members of the community have with the outer world – locally and in a wider European perspective? And how did these links – or missing links – influence the daily doings, the business, and the mentality of the brothers who led their lives on this island? Or did the Utstein community find itself in a situation of both-and, i.e., a state of dichotomy between isolation and integration?

A place with deep historical roots

At the time when the monastery at Utstein was founded in the 12th or 13th century – see the discussion below about the dating – the chosen site innermost in the bay of Klostervågen was already a place with a long history. From times immemorial around what we could call the birth of Norway some 10 000 years ago, when the last glaciers gradually withdrew, the coastline of this area is exactly the place where the oldest traces of human presence in this country have been found.[1] The huge field of stone-age rock carvings at the neighbouring Austre Åmøy bears witness to permanent settle-

1 Mari Høgestøl, *Arkeologiske undersøkelser i Rennesøy kommune, Rogaland, Sørvest-Norge* (Stavanger: Arkeologisk Museum, 1995), 95-6.

ment in the area based on agriculture and fishing. Archaeological traces from the Iron Age in the immediate neighbourhood of today's buildings offer evidence of the way of life and subsistence of the people living on this island in this period: a farmstead on the hill of Kneberfjellet west of the monastery, some 120 grave mounds, and a huge stone enclosure formed as a labyrinth on the hill of Bakkahodnet to the east, dated to the period AD 300-500. The enclosure may have been used as a cattle pen or for religious purposes.[2] Remains of iron production have been found, which most likely was not meant only for local use.

What is more relevant for the topic of this book, however, is the discovery at Utstein of a Roman bronze kettle, one of many similar kettles from the west coast of Norway known to have been produced in the Roman settlements along the Rhine in the two hundred years prior to the migration period.[3] We have no solid information about how these and other items from the Roman Empire, such as glass and warrior equipment, found their way to these coasts, but we know that an army from southwestern Norway made an unsuccessful attempt to raid southern Jutland in Roman times. These products may have arrived as a result of trade, booty from raids or as rewards for services as assisting soldiers – so-called *auxiliarii* – in the Roman army in the Netherlands or British Isles.[4] However this might have occurred, these finds enable us to conclude that the Skagerac and the North Sea waters were no obstacle for the people along this coastline to keeping in contact with at least the northern areas of the continent and the British Isles.

This leads to the next general point to be kept in mind in the following. From the migration period through the Viking age until the beginning of the Norwegian Middle Ages around the year 1000 the sea was an easier way of communication than was land.[5] In fact, the very name of Norway originates from the need to describe the old sailing and trade route along the western coast of the country up as far north as it was possible to come. This means that being located on islands along this coast did not necessarily imply social isolation, perhaps rather the contrary.

---

2 Jan Petersen, "Utgravninger på Utstein Kloster," *Stavanger Museum Årshefte* (Stavanger: Stavanger Museum, 1941-42); Eivind Heldaas Seland, "Utstein før Harald Hårfagre," in *Utstein Kloster – og Klosterøys historie*, ed. Eldbjørg Haug (Utstein: Stiftelsen Utstein Kloster, 2005), 21-9.

3 Odmund Møllerop, "Fortidsminner på Utstein," *Frá haug ok heidni* 1962, No 2; Seland, "Utstein," 41-2.

4 Torstein Jørgensen, "From Wessex to Western Norway: Some Perspectives on One Channel for the Christianisation Process," in *Church and People in Britain and Scandinavia*, ed. Ingmar Brohed (Lund: Lund University Press, 1996), 30: Per Hernæs, "Tidlige kristne impulser i Rogalands vikingtid," *Frá haug ok heidni* 1994, No. 1: 17-22.

5 Bjørn Myhre, "The Beginning of the Viking Age: Some Current Archaeological Problems," in *Viking Revaluations: Viking Society Centenary Symposium 14-15 May 1992*, ed. Anthony Faulkes and Richard Perkins (London: Viking Society for Northern Research, 1993), 182-205.

It would be beyond the scope of this study to go into detail about the development of the settlement at Utstein from the Viking Age until the founding of the monastery, a period when archaeological evidence is amplified with some, if not many, written sources. There is every indication that the site became a royal property and seat as early as the first unification of Norway into one kingdom in the late ninth century. After the battle of Hafrsfjord, near Stavanger, when Harald Fairhair gained sovereignty over a joint Norwegian kingdom, the sources tell that Utstein was one of the places where he often stayed.[6] With some smaller interruptions Utstein seems to have been one of several royal estates right up to the time when the land was donated to the Church as the site for a monastery foundation. Utstein was an excellent location for a king or a royal representative who wanted to control the traffic up and down the Norwegian west coast with its free sight to the sailing channel and its safe harbor with no less than three exits for attack or escape.

The founding of a monastery at Utstein

The time when this took place has been disputed since the source material on this point is scarce and insufficient. The traditional opinion that Utstein Monastery was founded by King Magnus the Lawmender in 1263/64 was launched by Christian Lange as early as 1847.[7] There is clear evidence in the sources that the community at Utstein belonged to the order of Augustinian canons.[8] This is, however, a fairly late date for the founding of an Augustinian canon monastery[9] since all the other foundations of the same order in the country were established in the twelfth century.[10] In recent times, Eldbjørg Haug has suggested that Utstein Monastery was established a century before, in 1164.[11] This was the year of King Magnus Erlingsson's coronation, on which occasion the father of the new underage king donated a site and land for the founding of the

---

6 Eldbjørg Haug, "Kongsgårdtid," in *Utstein Kloster – Klosterøys historie*, 56-61, 77.

7 Christian C. A. Lange, *De Norske Klostres Historie i Middelalderen*, 2nd rev. ed. (Christiania: Chr. Tønsbergs Forlag: 1856), 7, Oluf Kolsrud, "Stavanger bispestol," in *Stavanger 1125–1425–1925*, ed. Jan Petersen (Stavanger: Dreyer, 1925), 48-9, Jan Hendrich Lexow, "Utstein kloster i middelalderen" in *Foreningen til norske fortidsminnesmerkers bevarings årbok* 1987: 157.

8 *Diplomatarium Norvegicum* (henceforth *DN*) IV, no. 16.

9 The Augustinian canons are not to be confused with the mendicant Augustinian hermits, who were united into one order by Pope Alexander IV in 1256 with the installation of a prior general of the new order. The Augustinian hermits are normally referred to as Augustinians. No monastery of this latter order is known to have been established in Norway.

10 The Augustinian canon convent of St. John in Bergen may, however, have been founded already around or shortly after the mid-12th century. Christopher Hohler, "The Cathedral of St. Swithun in Stavanger in the Twelfth Century," *Journal of the British Archaeological Association*, Third Ser., 27 (1964): 95-6., Knut Helle, "Stavanger by og Utstein kloster," *Historisk Tidsskrift* 87, No 4 (2008): 594.

11 Eldbjørg Haug, "Klosteret på Utstein," in *Utstein Kloster – Klosterøys historie*, 119-21.

Augustinian Halsnøy Monastery on an island some 100 kilometres further north from Utstein. Haug's theory has been strongly opposed by Knut Helle.[12] There is, however, no indisputable documentation to support either of the two datings, but, as will be developed in the following, I find the mid-13th-century dating more likely.

Three entangled institutions

Another complicated issue that is closely linked to the dating and can be solved only by conjecture is the connection between Utstein Monastery, the St. Olav Monastery in the city of Stavanger, and the cathedral chapter in the same city. The way in which one interprets the interrelations between these three institutions also has a clear bearing upon the question of whether Utstein Monastery was an island on an island or not.

Again, the main problem is the overall scarcity of preserved sources. Lack of source material is by no means an exception for someone trying to investigate the medieval church of Norway since most of the ecclesiastical archival material was destroyed in connection with the Lutheran reformation. Some sources do exist, however, and it is possible to draw conclusions from sources touching on this topic more indirectly and also from more general knowledge. According to the sources, the three units must have existed as separate entities with their own names, but at the same time there is clear evidence that their operations, properties, and personnel were intertwined and overlapping.

The decision to establish chapters of canons at the Norwegian cathedrals was made in 1153, at the same time as the Norwegian archiepiscopal see in Nidaros was founded. Considering the generally high activity around the Stavanger cathedral and the kind of well-educated bishops available to hold the seat at the time there is every indication to suggest that the chapter commenced operations soon after. The chapter of Stavanger was established with twelve canons, although the community was not complete in number for shorter or longer periods. The first mentions of the St. Olav Monastery date from 1233 and 1236,[13] but the monastery must already have been in operation for a while at that time. Ludvig Daae found that there were grounds to believe that the monastery was established by Bishop Eirik Ivarson of Stavanger, a former scholar from the famous Augustinian canon monastery of St. Victor in Paris, some time during his Stavanger service between 1170 and 1188.[14] Edvard Bull, senior, suggests that the monas-

12 Helle, "Stavanger by og Utstein kloster," 577-605.

13 Referred to in Archbishop Henrik Kalteisen's Copy Book from the mid-fifteenth century, *Erkebiskop Henrik Kalteisens Kopibog*, ed. Alexander Bugge (Christiania: Kildeskriftfondet, 1899), 191-2.

14 Ludvig Daae, "En krønike om erkebiskopperne i Nidaros," in *Festskrift udgivet i Anledning af Trondhjems 900 Aars Jubilæum 1897* (Trondhjem: Selskabet, 1897), 9-10, 24-33, 119-120.

tery was founded before 1160, basing his theory on the legendary story of the miracles of St. Cuthbert. This legend tells about a young boy from Stavanger who was healed by the saint in 1172 after seven years suffering from an incurable illness. Before that, it is said about the boy that he had been a student in a Stavanger monastery for five years.[15] Whereas Daae was of the opinion that the St. Olav Monastery was an Augustinian canon foundation, Bull, followed by Anton W. Brøgger[16] and Oluf Kolsrud,[17] found it more likely that it belonged to the Benedictine order. The latter two maintain that the monastery had been founded by the first bishop of Stavanger, Reinald (1125-1138), himself regarded as a Benedictine monk from Winchester, who may have brought some brethren with him for a monastery foundation in Stavanger. If this is correct, the St. Olav Monastery dates back to before 1135. Most scholars after World War II, however, have tended to conclude that the St. Olav Monastery was an Augustinian canon monastery,[18] but the view that it belonged to the Benedictine order has also had its advocates.[19]

Eldbjørg Haug has lately launched the idea that the monastery may even have been founded during the reign of King Magnus Barefoot around the year 1100, that because it was an already-established Benedictine community was a reason why a bishopric could be established in Stavanger, and that the monks were responsible for the clerical services in the cathedral.[20] I find this early dating rather speculative, but the latter element about the convent's service in the cathedral is interesting, especially if St. Olav Monastery is imagined to have been founded as an Augustinian canon community after 1153. From elsewhere we know that Augustinian canons could be in full charge of or share in the duties of cathedral chapters. At the metropolitan see of Nidaros, for instance, the cathedral chapter consisted of both secular canons and regular canons from the Augustinian Helgeseter monastery.

15 Edvard Bull, "Et kloster i Stavanger," in *Stavanger Museums Årsskrift* 1910: 3-5, *Reginaldi monachi Dunelmensis Libellus de admirandis beati Cuthberti virtutibus*, ed. James Raine, Surtees Society, vol. 1 (London: J. B. Nichols and son, 1835), 249.

16 Anton Wilhelm Brøgger, *Stavangers historie i middelalderen* (Stavanger: Dreyer, 1915).

17 Kolsrud, "Stavanger bispestol," 41. Kolsrud later changed his opinion in *Noregs Kyrkjesoga I: Millomalderen* (Oslo: Aschehoug, 1958), 206-7.

18 Arne Odd Johnsen, *Studier vedrørende kardinal Nicolaus Brekespears legasjontil Norden* (Oslo: Fabritius, 1945), 117-8, Eirik Vandvik, *Latinske dokument til norsk historie* (Oslo: Samlaget, 1959), 24-5, 185-6, Asgaut Steinnes, "Klostergods i Rogaland," *Ætt og Heim* 1967: 7-20, Knut Helle, "*Under kirke og kongemakt 1130-1350*," in *Aschehougs Norgeshistorie*, ed. idem et al., vol. 3 (Oslo: Aschehoug, 1995), 119, 165, Erik Gunnes, *Erkebiskop Øystein: Stasmann og kirkebygger* (Oslo: Aschehoug, 1996), 194, Tore. Nyberg, *Monasticism in North-Western Europe – 800-1200* (Aldershot: Ashgate, 2000), 98, 154, 222-4.

19 Olav Nenseter, "Å lære gjennom ord og eksempel: Augustinerklostrene på Vestlandets religiose funksjoner," BA thesis (University of Oslo, 2002), 34-6.

20 Haug, "Klosteret på Utstein," 119-21.

Some of the details of the intertwined connection between these three units bear looking at as they appear in the sources. First, one has to take into account that St. Olav Monastery is specifically mentioned in only one preserved document besides the texts from 1233 and 1236 mentioned above. This dates from 1478. Bishop Alv Thorgardssøn states in his will from this year that he favors St. Olav Monastery with one barrel of tar.[21] From this, one can deduce that a reference to a monastery of this name must have been meaningful as late as the late 15th century. In the same will the abbot of Utstein is given a black fur coat. The monastery *church* of St. Olav is, however, mentioned in several documents. Another interesting piece of information is given in a document from 1295 stating that Trond and Ivar, two Augustinian brethren at Utstein, are responsible for the clerical service in the St. Olav Church.[22] And when it comes to properties, much of the land around the St. Olav Church belonged to Utstein Monastery[23] and a mill in the city is known to have been the joint property of both.[24]

The general, but still somewhat blurred, picture that appears from these scattered pieces of information is that an Augustinian canon monastery dedicated to St. Olav was established in Stavanger in the wake of the founding of the Norwegian metropolitan see in 1153. There is also good reason to suggest, but not to conclude, that at least some of the brethren had a share in the cathedral chapter together with secular canons. Whereas St. Olav monastery is mentioned as a monastery with both an abbot and a prior in 1236 – the two are listed as witnesses to a grant of 30 mark silver to the archbishop[25] – it is most striking that neither leaders nor members of a St. Olav community are mentioned in any later sources. In contrast, both the abbot of Utstein and some of the monks there appear in several texts.[26] Somehow, the founding of a monastery at Utstein in the 1260s seems to have taken place at the expense of the St. Olav convent, and possibly it may simply have been absorbed by the new, larger, and probably more prestigious community at Utstein as part of its foundation – with the donation of a royal estate as another cornerstone. These circumstances surrounding the property, and the Utstein brethren doing service at the St. Olav Church, fit well with this theory. The barrel of tar for St. Olav Monastery, mentioned in the will of Bishop Alv in 1478, may simply have been a necessary grant for preserving the wooden parts of

---

21 *DN* IV, no. 987.

22 *DN* IV, no. 16.

23 *DN* II, no. 134.

24 *DN* IV, no. 537.

25 *Datum stawangrie anno domini mccxxxvj sub his testibus domino S nidrosiensi archiepiscopo, domino askillo stawangrensi, domino N grenlandensi, N abate de sancte olauo apud stawangream, E priore eiusdem loci, berone et augustino et torgaro canonicis nidrosiensibus.*" *Erkebiskop Henrik Kalteisens Kopibog*, 192.

26 See, for instance, *DN* I, no. 1047; II, no. 77, 134; IV, no. 16, 17, 55, 145, 212, 214, 215, 403, 537, 863, 892, 1094; VII, no 539, 540; IX, no 133; XII, no. 19; XIII, no. 44; XV, No. 531; XVII, no. 188, 430.

a building in decay still generally referred to as St. Olav Monastery in spite of the monastery no longer being there.

The fact that the two convents were dedicated to different patron saints, St. Olav and St. Laurentius, confronts us with a problem and may be interpreted as a sign of discontinuity between the two, but the problem is by no means insuperable. The mid-12th century was a time when the St. Olav cult was at its peak in Norway. On the solemn occasion of the founding of the Norwegian archiepiscopal see in 1153 a long praise poem, *Geisli*, "the Sunbeam," was presented in honour of St. Olav. And when Magnus Erlingsson was crowned in 1164, his crown was sacrificed on St. Olav's altar and his reign placed under the protection of St. Olav, who was called *rex perpetuus norvegiae*. For a monastery to be established in Stavanger at this time, with no institutional representation of the national saint previously, it was but natural to dedicate this community to St. Olav. A hundred years later, when the monastery at Utstein seems to have been founded, the St. Olav community house in Stavanger, and even more the church which was part of it, must have been well established and combined with these institutions, which there was no reason to alter. There is no written evidence for why St. Laurentius was chosen as the patron saint of Utstein apart from the fact that he was a well known and popular saint of the time.

If this reasoning, with all its uncertainties, describes the main line of development of the connection between the two monasteries, the Utstein Monastery appears, as far as the focus of this book is concerned, as an institution with strong links to the mainland. These links consisted of duties in both places by the Utstein personnel, of both the abbot and some of the brethren employed by the higher Norwegian clergy and even by the pope for special commissions, of property connections both to the city of Stavanger and elsewhere, and of the ideological impulses of the Augustinian canon order in more general terms.

When it comes to the relation between Utstein Monastery and the cathedral chapter in Stavanger, the picture becomes more intricate, but some indications do exist. The overall tendency of the sources points in the direction of two separate units. One element is the fact that in all places where the canons of the chapter and their economic sustenance are mentioned it consists of prebendaries and land of their own ownership.[27] Properties could also be connected to the different altars of the cathedral at which each canon performed his services. If this was the case without exception, the chapter must have consisted merely of secular canons, since regular canons were catered for by their convent. A possible solution is that when the Augustinian canons of St. Olav, some of whom may have served as regular canons in the cathedral chapter, moved to Utstein in the 1260s they were replaced by a secular chapter.

---

27 For instance, *DN* XVII, no. 188.

Another indication pointing in the same direction is expressed by the fact that the abbot of Utstein could be used as an independent figure in the jurisdictional affairs of the bishop of Stavanger. The actual matter was a fierce conflict between the bishop and the cathedral chapter, in fact, one of the best documented pieces of Norwegian medieval church history. In a diploma from 1297, the abbot, together with two local priests, is instructed by the bishop of Oslo to summon the bishop of Stavanger to the ecclesiastical court in Oslo.[28] If some of the abbot's subordinate monks at this point in time were members of the cathedral chapter, which had a part in the conflict, the abbot would have been prejudiced and not likely to be called upon for such a commission.

On the other hand the demarcation line between the two units may not have been that sharp all the time. At least, there is one example of one and the same person first referred to as canon of the cathedral chapter and later as abbot of Utstein. In 1525 the canon of the cathedral, Torgils Amundssøn, was the sender of a penitentiary supplication to the pope asking absolution for having killed another man.[29] Five years later the same Torgils is in office as abbot of Utstein.[30] Torgils may, of course, have been a secular canon in 1525, joined the Augustinian convent at Utstein, made a rapid career there and become abbot. There is also the possibility, however, that he already served in the cathedral chapter as a regular canon in 1525.

Some glimpses of the life at Utstein

When it comes to the question of daily life at Utstein Monastery and the links of the community with the outer world, we unfortunately have only small pieces of information. Like other Augustinian canon communities, Utstein recruited its brethren from the relatively well-educated aristocracy. As an important land owner the monastery at Utstein must have had its main income from its estates.[31] There is no room here to go into details of this. It also possessed fishing rights, such as for salmon in the Figgjo River.[32] All this means that the monastery played an important part and was an integral factor in the economic structure and activity of the region; being centered on an island was no obstacle.

The trading network of Utstein also seems to have had a wide outreach. Early 14th-century records from the English port of King's Lynn in Norfolk both mention a

---

28 *DN* IV, no. 17.

29 *Archivio Segreto Vaticano, Penitenzieria Apostolica, Reg. Matrim et Div.* vol. 73, fol. 1135r-1136r, Torstein Jørgensen and Gastone Saletnich, *Synder og Pavemakt: Botsbrev fra Den Norske Kirkeprovins og Suderøyene til Pavestolen 1438-1531* (Stavanger: Misjonshøgskolens forlag, 2004), 99-100, 165-6.

30 *DN* IV, no. 1094.

31 Helle, "Stavanger by og Utstein kloster," 599.

32 *DN* IV, no. 1075.

St. Laurentii busse and an *Utstein busse.*[33] *Busse* was a common term for a quite large, flat bottomed cargo ship. Most likely the St. Laurentii and the Utstein busse are one and the same ship since they are recorded as having the same captain, one Tor Gaut, in the same year of 1303. There should be no doubt that the ship was a freighter owned by the Utstein monastery. Listed cargoes from the ship are timber, stock fish, honey, wheat, conifer sprouts and bresil, which was a yellowish dyestuff. Most likely the Utstein busse carried cargoes from the monastery's own production or trading, but independent merchants could also have rented space on the busse for their own export items. To what extent shipping activity was important in the total economy of Utstein and how long it lasted is impossible to say, but the little glimpse that the King's Lynn records offer fills in a picture of a center with several outreaching links.

Both for the Stavanger cathedral chapter and for Utstein episodes of conflict and strife are better documented than times when life went on in its normal track. In the case of Utstein two intensive fights between the abbot of Utstein and the bishop of Stavanger are reported in the sources. The former dates from the years 1333 to 1341,[34] the latter is evidenced in texts from 1515.[35]

On the administrative level the former dispute between the two, both sharing the name of Eirik, dealt with the right to inspect the non-exempt monastery by a visitation. When arriving with his retinue the bishop finds the doors of the monastery locked. He breaks them open by force and celebrates mass in the monastery church. In short, the conflict results in both the abbot and the bishop accusing one another by letters of complaint, written by each of them and addressed to the pope. The abbot explains that the reason why he did not allow the bishop access to the monastery was that the latter had been excommunicated by the papal nuncio, Pierre Gervais, because of failing to pay the six-years tithe, and that he was unable to communicate with the bishop in a normal way since the latter had threatened to imprison him in the tower of the cathedral. The bishop, for his part, accuses the abbot of disrespect for his authority and, allegedly on the basis of complaints from the Utstein friars, of nightly boozing and sex with a noble lady, of theft of and waste of the convent treasures, of forcing the confessor of the friars to reveal their confessed sins to the abbot, followed by punishment, and of an immensely brutal rule. As an example of his brutality the abbot falsely accused one of the friars of having stolen a robe and put him in bolts and chains for four days and three nights without food and drink with the result that he nearly died. Our knowledge of how the conflict was solved is insufficient and based only on a short 1341 diploma from Bishop Håkon of Bergen to Bishop Eirik of Stavanger, referring to a now-lost

---

33 *DN* XIX, no. 422, 464, 467. Eldbjørg Haug, "Godseier og skipsreder," in *Utstein Kloster – og Klosterøys historie*, 166-7.

34 *DN* IV, no. 212, 214, 215. See also Lange, *De Norske Klostres Historie*, 379-82.

35 *DN* IV, no. 539, 540. See also Lange, *De Norske Klostres Historie*, 384-6.

papal letter in which the Holy Father somewhat surprisingly seems to have decided in favour of the abbot.

The second recorded conflict, occurring some two hundred years later, is of a similar kind. The main issues were again the right of the bishop to visitation and the duty of the abbot to pay tribute to the see. This time the bishop took more drastic measures. On an occasion when the then-Abbot Henrik was absent on a journey to Copenhagen, Bishop Hoskold sent some of his men to the monastery, who wasted its provisions, listed its valuables, and took away some valuable items. A year later the bishop himself appeared at night at Utstein with a group of armed men, broke down the door, put ladders to the windows and the tower, and entered the monastery. The abbot, whom they found in bed, was beaten with a war hammer and brought, wounded and bleeding, to the prison in the cathedral tower. On the same occasion the bishop also stole a gold ring from the abbot as well as the cross he wore. After 100 days of imprisonment, the abbot, as he himself states, with the help of the Virgin Mary and St. Anne, managed to escape, and under the protection of two royal representatives travelled to Copenhagen and presented his complaints to the king. The case cannot have been solved immediately, since in 1524 it is documented to have been forwarded to a meeting of the Norwegian National Council in the same year.

These two conflicts render a somewhat bizarre illustration of Utstein's links to the surrounding world, namely that of violent skirmishes, *in casu* between the abbot and the bishop of Stavanger. Both conflicts were also reported to and dealt with by higher ecclesiastical and civil authorities.

Conclusion

In geographic terms Utstein Monastery was definitely an insular institution. The Latin word *insula* has the modern derivates of insulation and isolation, the latter via the Italian term *isola*, showing that island and isolation appear as notions closely related philologically. Surrounded by water, islands, of course, did have a natural fence around them that gave them a most visible kind of physical isolation, but in a country of steep hills and mountains, wild forests, and with more than enough wind, good sailing vessels stood out as the easiest means of transport. Boats could carry heavy loads as well as people in quite large numbers, and with good winds they could keep up high speed with the use of limited manpower. In such an environment islands and coastal settlements could, if strategically located, not only be easily accessible, but become centers of trade and industry, administration, and social life in general. Utstein Monastery never developed into such a large center. But this presentation should have shown that the community behind the convent walls at Utstein in a number of respects did not lead an isolated existence as an island on an island. Still, for the learned brethren in their daily duties of regulated prayers and mass, reading and writing, Utstein must have been quite a peaceful place. But this was probably more thanks to the monastery walls than to the surrounding sea.

LIST OF CONTRIBUTORS

Margaret Elphinstone, Department of English Studies, University of Strathclyde, Glasgow, Scotland (retired)

Eldar Heide, Centre for Medieval Studies, University of Bergen, P.O. BOX 7805, 5020 Bergen, Norway

Johnny Grandjean Gøgsig Jacobsen, Department of Scandinavian Research, University of Copenhagen, Njalsgade 80, 2300 Copenhagen S, Denmark

Gerhard Jaritz, Department of Medieval Studies, Central European University, Nádor utca 9, 1051 Budapest, Hungary, and Institut für Realienkunde, Austrian Academy of Sciences, Körnermarkt 13, 3500 Krems, Austria

Torstein Jørgensen, Centre for Medieval Studies, University of Bergen, P.O. BOX 7805, 5020 Bergen, Norway, and School of Mission & Theology, Misjonsmarka 12, 4024 Stavanger, Norway

Juhan Kreem, Tallinn City Archives, Tolli 6, Tallinn, 10133, Estonia

Else Mundal, Centre for Medieval Studies, University of Bergen, P.O. BOX 7805, 5020 Bergen, Norway

Torben Kjersgaard Nielsen, Institute of History and International Studies, University of Aalborg, Fibigerstræde 5, 9220 Aalborg, Denmark

Felicitas Schmieder, Fernuniversität Hagen, Lehrgebiet Geschichte und Gegenwart Alteuropas, Universitätsstraße 11, TGZ, 58084 Hagen, Germany

Bernt Øyvind Thorvaldsen, Høgskulen i Volda, Joplassvegen 11, 6101 Volda, Norway

Kristel Zilmer, Centre for Medieval Studies, University of Bergen, P.O. BOX 7805, 5020 Bergen, Norway

INDEX

Absalon, Bishop, 36
Adam and Eve, 3, 4, 84
Aelred of Rievaulx, 49
Ahmad ibn Fadlan, 59, 61
Ailbe, s. Island of the Community of Ailbe
Äjjih, 73
Alexander, 113-6
Alexander the Great, 114, 117
Algirdas, Duke of Lithuania, 124
Angela of Foligno, 55
Ari Másson, 81-3, 86
Arresø, 41
Asserbo, 41
Atlas Catalan, 12, 13, 16, 17
Atlantis, 90
Augustinian ideas, 46
Augustinians, 36, 38, 43
Austre Åmøy, 128
Avalon, 63
Azores, 93
álfar, 64, 69
Ásmundr, 67
Baffin Land, 75
Bakkahodnet, 129
Baldr, 59, 72, 97
Barlaam and Josaphat, 112
Bárðar saga, 73, 75
bears, 114, 116
Beatrice of Nazareth, 54, 55
Benedict, Rule of Saint, 37
Benedictines, 38, 39, 41-3
Beowulf, 59, 65
Bergen, 41
  Bishop Håkon, 136
Bernard of Clairvaux, Saint, 50
Bestla, 103
Bianco, Andrea, 15, 17, 18
Bifrost, 106
Bifrǫst, rainbow bridge, 61
Bjarnar saga hítdœlakappa, 85
Bjǫrn breiðvíkingakappi, 82, 83, 86
Blå Jungfrun, 79
Boccaccio, Giovanni, 11
Borgia world map, 9, 10
Borr, 100, 103
Borre, 60
Bosjö, 38
Bǫðvarr, son of Egill Skallagrímsson, 67
Brasil, 17, 18
Brendan, Saint, 15-9, 83, 84, 90, 94, 95
bridges of the sky, 61
Brittia, 62
Brísingamen, see ring of the Brisings
Bunyan, 94
Bynum, Carolyn Walker, 54, 55
Canary Islands, 10-7, 91
Canterbury, 42
Cape Wrath, 93
Carthusians, 41
Catalan Estense world map, 8, 14

Cistercians, 37, 39, 40, 43
Columba, Saint, 90
Columbus, Christopher, 22
Constable, Giles, 52
Cresques Abraham and Jafuda, 12
Crusade, Baltic, 123
Cuthbert, Saint, 132
Dietrich's Flight, 116
Dominicans, 40-4
Dominic Loricatus, 52
Dorpat, 125
dragons, 114, 116
echtra, 83, 84
Edda, 61
    Younger, 77
    Prose, 97, 99, 104-6
Eden, 9, 19, 93
Egill Skallagrímsson, 67
Egils saga einhenda, 67, 77
einfœtingr, 85
Eiríksmál, 61
Eiríks saga rauða, 31, 81, 83, 85-7
Elfland, 92
elves, 69, s. also *álfar*, *háldit*
Enlightenment, Scottish, 89, 90
Eskil, Archbishop, 41
Eskilsø, 36, 38
Esrum, 40
Euphrates, 6
Evenks, 70
Exordium parvum, 40
Eyrbyggja saga, 66, 67, 77, 81-3
Faerie, 91, 92, 94
fairies, 70, 79, s. also *álfar*, *háldit*, *hulderfolket*, *mearrahaldit*
Falster, 38
Faroe Islands, 84
Fárbauti, 98, 101, 104
Fellin, 125, 126
Fensalir, 65
Figgio, River, 135
Finistère, 62
Finnur Jónsson, 102
Fjǫlsvinnsmál, 63
Flateyjarbók, 26, 104
Floki, 89
Folkvangr, 61
fornaldarsaga, 85
Four Rivers, 4, 5
Fǫr Skírnis, s. *Skírnismál*
Franciscans, 40, 41, 42, 43
Freyja, 61, 70, 97
Frigg, 61, 65
Fyn, 38
Ganges, 6
Garðarshólmr, 25
Gautreks saga, 61
Gavnø, 38, 40
Geirrøðr, 62
Gerðr, 107
Germain, Jean, 20, 22
Gertrude of Delft, 54
Gesta Danorum, 61
Gestr, 75
Ghilbert de Lannoy, 125
Gilbert Foliot, 52
Gjallarhorn, 105
Gjǫll, river, 59
Gladsaxe, 40
Godøya, 32
Goðeyjar, 32, 73
Gotland, 120
Gǫngu-Hrólfs saga, 77
grave islands, 60
Greenland, 81, 83, 85
Grendel, 65
Grettis saga, 65
Grímnismál, 61, 102

Grímr, 77
Grógaldr, 63
Grœnlendinga saga, 85-7
Gudfjelløya, 73
Guido of Ravenna, 11
Guigo II, prior of the Grande Chartreuse, 46, 47
Guta saga, 76
Gylfaginning, 79
Gyon, 6
Haddingus, 69
Hadewijch, 54
hafnýra, s. sea-kidney
Hafrsfjord, battle of, 130
Halsnøy Monastery, 131
Hanseatic League, 120, 122-5
Hapsal, 122, 123
Harald Fairhair, King of Norway, 130
Haraldr Gormsson, King of Denmark, 28, 29, 33
Harrison, John, 89
haugbrot, 77
Hauksbók, 106
Hákon, Earl, 27-30, 62
háldit, 76
Hárbarðsljóð, 31,62, 76
Hälsingland, 79
Hebrides, 92
Heimdallargaldr, 104
Heimdallr, 97-102, 104-9
Heimskringla, 29, 30, 32, 33
Hel, 59, 72
Helgafell motif, 77
Helgakviða Hundingsbana, 61
Helgasjön, 74
Helgeseter monastery, 132
Helgøya, 74, 75
Helgi, 61
Helgö, 74, 75, 79
Heligön, 74, 75, 79
Helluland, 75, 85
Helsingborg, 41
Henry of Livonia, 122, 124
Henry of Mainz, 4
Hereford world map, 6, 16
Hermóðr, 72
Hertig Silverdal, 63
Hervarar saga ok Heiðreks, 77
Hesperides, 90, 94
Himinbjǫrg, 105, 106, 108
Hjarðarholt, 97, 1025
Hjǫrungavágr, 26
Hlér, s. Ægir
Hlésey, s. Læsø
Hlymreka, s. Limerick
Hofvin, s. *Hovinsholm*
Holmen, 74, 75
Holy Island, 74, 79
Hordaland, 62
House Lay, 97, 100, 101, 105, 108-10
Hovinsholm, 74
Hǫlgabrúðr, 62
Hǫrðr, 67
Hǫrund, 26
Hǫrundarfjǫrðr, 26
Hrafn Hlymreksfari, 82
Hugh of Lincoln, 53, 54
hulderfolket, 63, 76
Hume, David, 89
Husdrapa, 97
Hvítramannaland, 81-7
Hy-Brazil, 90
Hyndluljóð (Song of Hyndla), 104, 107-9
imitatio Christi, 51, 53, 55
immram, 83, 84
Inari, Lake, 73
Indian islands, 3
Innocent III, Pope, 47

Insulae fortunatae, 10, 11, 13, 16
Irben, strait of, 125
Irpa, 62
Isidor of Seville, 6, 9, 11, 15, 16
Island of Birds, 94
Island of the Community of Ailbe, 94
Island of Sheep, 94
Írland it mikla, 81, 83, 85
Ívarr Víðfaðmi, legendary King of Sweden, 67
Jerusalem, heavenly, 20
Jerwen, 124
John of Marignola, 3, 6
Jómsvíkinga saga, 25, 28, 29, 62
Judas Iscariot, 84
Jylland, 41
Kalmarsund, 79
Kalvø, 39
Karlsfeni, 31, 81, 86, 87
Kattegat, sea of, 62
Ketils saga hængs, 65, 77
King's Lynn, Norfolk, 136
Klosterøy, 38, 128
Klostervågen, 128
Knapplundsøya, 32
Kneberfjellet, 129
Knut Holmgersson Långe, King of Sweden, 41
Kock, Reimar, 123
Konrad von Würzburg, 114, 116, 117
Lahepere, 123
Landnámabók, 24, 81, 82, 83, 86
Landtag, Livonian, 126
Lanzelet, 112
Laufey, 104
Laurentius, Saint, 134
Laxdœla saga, 66, 67, 97
Lemnos, 117
Lewis, 93
Lidwina of Schiedam, 54
Limerick, 82
Lindisfarne, 38
Linné, Carl von, 71
lions, 114
Lofoten, 77
Lokasenna, 31, 76, 106
Loki, 31, 97-102, 104, 106-9
Lolland, 38
London, 42
Lothar of Segni, s. Innocent III
Lucifer, 84
Lukardis of Oberweimar, 54
Luther, Martin, 9
Lutheran Reformation, 131`
Lübeck Chronicle, 123
Læsø, 31, 62, 67, 72, 76-9
Mabinogion, 94
Mag Mell, s. *Tír na n-Óg*
Magnus Barefoot, King of Norway, 132
Magnus Erlingsson, King of Norway, 130, 134
Magnus the Lawmender, King of Norway, 130
Malmö, 41
Mamluks, 7
Man, Isle of, 62, 93
Margaret, Queen of Denmark, 40
Margaret of Faenza, 54
Markland, 81, 85
Marvels, 115, 117
Mauretania, 13
Mauro, Fra, 17
Mälardalen, 42
mearrahåldit, 63, 76
Mellin, Ludwig August Graf, 126
Mickwitz, 125
Minch, 93

Mjøsa, Lake, 74
Modena map, 7, 16, 17
Monk's ladder, 46, 47
monsters, 114, 115
Moon, 124
Moonsund, 124-6
Mors, 29, 38
Moster, 34
Naglfar, 62
Nargen, 122
Narva, 121, 125
Navigatio sancti Brendani abbatis, 83, 84, 94, 95
Nenets, 70
Neo-Platonic ideas, 46, 55
Nidaros, archdiocese, 131, 132, 134
Nile, 6, 7
Nio, 117
Noah, 89
noaidi, 66
Nordfjord, 34
Novgorod, 125
Nuckö, 122
Næstved, 38
Oddr Snorrason, 32-4
Odensjön, Lake, 67, 68
Olav, Saint, 75, 134
Onega, Lake, 76
Orkneys, 82, 83, 93
Osilians, 122, 123
Oslo, bishop of, 135
Oswald von Wolkenstein, 117
Otherworld, 57-80, 83
Oxford, 42
Óðinn, 30, 31, 57, 61, 63, 65-8, 70, 76, 77, 79, 97, 99, 100, 101-7
Óláfr Haraldsson, King of Norway, 33
Óláfr Hǫskuldsson, 97
Óláfr Tryggvason, King of Norway, 32-4
Óláfs saga Tryggvasonar, 32-4
Óláfs saga Tryggvasonar en mesta, 28, 29, 33
Ösel, 124, 125
bishop of, 126
Øm, 39
Pakasaivo, Lake, 68
Paradise, 3-22, 36, 37, 61, 63, 84
Parceval, 114
Paris
Augustinian canon monastery of St. Victor, 131
Partonopier and Meliur, 116
Peipus, Lake, 124, 125
penetratio Christi, 53
Penitentiary, Apostolic, 135
Perdita, 14, 15
Pernau, 121, 123, 125
Peter of Celle, Bishop of Chartres, 50, 51
Peter Damian, 51, 52
Philip of Burgundy, Duke, 20
Phison, 6
Pierre Gervais, papal nuncio, 136
Pilgrim's Progress, 94
Polus antarcticus, 113
Poppo, Bishop, 29, 30
Porkala, 121
Premonstratensians, 41, 42
Prester John, 8
Pretanicca, Pytheas', 92, 93
Prímsignd, 26-8, 62
Procopius, 59, 62
Prose-Lancelot, 114-6
Psalms, 52
Pskov, 125
pyhä, 78

Pytheas, 90, 92, 93
Ragnarǫk, 107
Ragnars saga loðbrókar, 77
Rathlin Island, 92
Rán, 62, 66, 67
Reid, Thomas, 89
Renner, Johann, 124
Reval, 121-5
Reyðarfjörður, 77
Ribe, 44
Riga, 121, 127
ring of the Brisings, 97
Rome, 44
Roskilde, 38
Rudolf von Ems, 112, 114, 115
Rum, 92
Runö, 127
Rupert of Deutz, 53
Russow, Balthasar, 123
Rügen, 38
saajve, 66, 67, 70
Saami mythology, 63, 66, 68, 70, 73-7
Salis, 124
Saltfjorden, 32
Saltstraumen, 32, 73
Sandflesa, 76
Samsey, s. Samsø
Samsø, 31, 76-8
Saramego, José, 88, 95, 96
savagery, 113, 114, 117
Saxo Grammaticus, 61, 69
Sága, 65
Schildau, 124
Scilly Isles, 33
sea-kidney, 97, 99, 101, 102, 109
Seifrit, 113-6, 118
Seley, 77, 79
Selja, 34
Severin, Tim, 95
Sigtuna, 41
Silvestri Domenico, 10-7
Singasteinn, 97-99, 101, 102
Sinfjǫtli, 61
sjógýgr, 77
Sjælland, 38, 40
Skaldskaparmál, 77
Skírnir, 107
Skírnismál, 107
Skjerstadfjorden, 32
Skjǫldunga saga, 67
Sko, 41
skrælingar, 81
Skrova, 77, 79
Skåne, 40, 41
sledge-boats, 127
sledgeway, 125, 126
Snorra Edda, 25
Snorri Sturluson, 65, 77, 79, 80, 97, 104, 106
Snorri góði, chieftain, 82
Sonatorrek, 66
Son of the Sun, 63
Sontagana, 124
Sorø, 39
Sǫgubrot af fornkonungum, 67
Sreva, River, 124
stabilitas loci, 37, 42
stag, 116
Starkaðr, 61, 62
Stavanger
  Bishop Alv Thorgardssøn, 133
  Bishop Eirik, 136
  Bishop Eirik Ivarson, 131
  Bishop Hoskold, 137
  Bishop Reinald
  St: Olav's monastery, 131-4
Stevenson family of engineers, 89
Stockholm, 123

Storjungfrun, 79
Strabo, 90, 93
Straumsey, 31
sub-sea fairy world, 68
subterraneans, s. *underjordiska*
Søkkvabekkr, 65
Sula, 26
Sunnhordland, 34
Svávarsson, Garðarr, 25
Sævarstǫð, 31
Tacitus, 62
Teutonic Knights, 124
thing of trolls, s. *trollaþing*
Thule, Ultima, 93, 110
Tigris, 6
Tír fo Thuinn, s. *Tír na n-Óg*
Tír na n-Óg, 63, 64, 69, 76, 78, 90, 94
Tjåehkere, 73
Torkö, 40
Trojan War, 114, 116, 117
trollaþing, 65
Trolls, 85
Trondheim, 75
True Thomas, 91, 92, 94
Tunnsjøen, Lake, 73
Turkic crusade, 20
Ukonsaari, 73
Ulrich von Etzenbach, 117
Ulrich von Zatzikhoven, 112, 113
underjordiska, 70
Ungen Svejdal, 63
Unknown Island, 88-96
Utrøst, 76
Utstein, 38, 126-36
Abbot Eirik, 136
Abbot Henrik, 137
Abbot Torgils Amundssøn, 135
Úlfr Uggason, 97, 99, 100-2, 105, 109
Vafþrúðnismál, 62
Valhǫll, 61, 63, 67
Vágasker, 97
Velletri world map, 9, 10
Veng, 39
Vigra, 26
Vikings, 59, 84
Vita Brendani, 84
Vínland, 31,81, 83, 85, 86
Vínland it góða, 82
Volga, 59
Vǫlundarkviða, 31
Vǫlundr, 31
Vǫluspá, 86, 99-101106, 107
Walsperger, Andreas, 20, 21
waterway, 123, 124
wildness, 113-5, 117
William of Æbelholt,, 36, 37
Wismar, 123
Witzjärv, Lake, 124
Worms, 122
Wulfric, Saint, 48
Yggdrasil, 94
Ymir, 102-4
Þangbrandr., 109
Þorfinnr, Earl of the Orkneys, 82, 83
Þorgerðr, 62
Þorkell Gellisson, 82
Þorsteins þáttr bǿjarmagns, 64, 69
Þorvaldr veili, 108
Þórr, 31, 62, 72, 76, 97
Þuríð, sister of chieftain Snorri góði, 82
Æbelholt, 38
Ægir, 62, 66, 76, 106
Århus, 43, 44